Ubiquitous Technologies for Emotion Recognition

Ubiquitous Technologies for Emotion Recognition

Editors

Oresti Banos
Luis A. Castro
Claudia Villalonga

MDPI • Basel • Beijing • Wuhan • Barcelona • Belgrade • Manchester • Tokyo • Cluj • Tianjin

Editors
Oresti Banos
Department of Computer Architecture and Computer Technology
University of Granada
Granada
Spain

Luis A. Castro
Dept. of Computing and Design
Sonora Institute of Technology
Ciudad Obregón
Mexico

Claudia Villalonga
Department of Computer Architecture and Computer Technology
University of Granada
Granada
Spain

Editorial Office
MDPI
St. Alban-Anlage 66
4052 Basel, Switzerland

This is a reprint of articles from the Special Issue published online in the open access journal *Applied Sciences* (ISSN 2076-3417) (available at: www.mdpi.com/journal/applsci/special_issues/Emotion_Recog).

For citation purposes, cite each article independently as indicated on the article page online and as indicated below:

LastName, A.A.; LastName, B.B.; LastName, C.C. Article Title. *Journal Name* **Year**, *Volume Number*, Page Range.

ISBN 978-3-0365-1802-2 (Hbk)
ISBN 978-3-0365-1801-5 (PDF)

Contents

About the Editors

Oresti Banos

Oresti Baños is a Tenured Professor of Computational Behaviour Modelling at the University of Granada (Spain, 2019–present). He is also a Senior Research Scientist affiliated with the Centre for Information and Communication Technologies of the University of Granada (CITIC-UGR). He is a Research Collaborator at Kyung Hee University (South Korea, 2016–present) and the University of Twente (Netherlands, 2018–present). He is a former Assistant Professor at the University of Twente (2016–2018), Postdoctoral Fellow at Kyung Hee University (2014–2016), Predoctoral Fellow at CITIC-UGR (2010–2014), and Visiting Scholar at the Technical University of Eindhoven (Netherlands, 2012), the Swiss Federal Institute of Technology Zurich (Switzerland, 2011), and the University of Alabama (USA, 2011). His research focuses on the intersection of wearable, ubiquitous, and mobile computing with data mining and artificial intelligence for digital health and wellness applications.

Luis A. Castro

Luis A. Castro works as a full professor at the Department of Computing and Design at the Sonora Institute of Technology (ITSON), México. He holds a PhD in Informatics from the University of Manchester, UK. Castro's main research interests are community informatics, intelligent systems; human–computer interactions and interaction design, and ubiquitous and mobile computing. Dr. Castro is a professional member of the ACM, and a member of the National System of Researchers from the National Council for Science and Technology in Mexico (SNI-CONACYT). He is the former president of the Mexican Association on Human–Computer Interaction (AMexIHC).

Claudia Villalonga

Claudia Villalonga holds a PhD in Information and Communication Technologies from the Universidad de Granada, Spain, a Master's Degree in Business Management (Administration of Organizations in the Knowledge Economy) from the Universitat Oberta de Catalunya, Spain, and a Master's Degree in Telecommunications Engineering from the Universitat Politècnica de Catalunya, Spain. She is currently a lecturer at the Universidad de Granada, Spain, and the former Director of the Master's Degree in Artificial Intelligence at the Universidad Internacional de La Rioja, Spain. Claudia Villalonga has worked in international environments for the research and innovation centers of several major multinational enterprises: CGI in Spain, SAP in Switzerland, and NEC in Germany. She has worked in several academic institutions: the Swiss Federal Institute of Technology Zurich (ETHZ) in Switzerland, Kyung Hee University (KHU) in South Korea, and the University of Twente in the Netherlands.

Preface to "Ubiquitous Technologies for Emotion Recognition"

Emotions play a very important role in how we think and behave. As such, the emotions we feel every day can compel us to act and influence the decisions and plans we make about our lives. Being able to measure, analyze, and better comprehend how or why our emotions may change is thus of much relevance to understand human behavior and its consequences. Despite the great efforts made in the past in the study of human emotions, it is only now, with the advent of wearable, mobile, and ubiquitous technologies, that we can aim to sense and recognize emotions, continuously and in real time. This book brings together the latest experiences, findings, and developments regarding ubiquitous sensing, modeling, and the recognition of human emotions.

Oresti Banos, Luis A. Castro, Claudia Villalonga
Editors

applied sciences

MDPI

Editorial

Ubiquitous Technologies for Emotion Recognition

Oresti Banos [1,*,†,‡], Luis A. Castro [2,‡] and Claudia Villalonga [1,‡]

[1] Department of Computer Architecture and Technology, CITIC, University of Granada, 18014 Granada, Spain; cvillalonga@ugr.es
[2] Department of Computing and Design, Sonora Institute of Technology (ITSON), Ciudad Obregon 85130, Mexico; luis.castro@acm.org
* Correspondence: oresti@ugr.es
† Current address: C/Periodista Rafael Gómez Montero 2, 18071 Granada, Spain.
‡ These authors contributed equally to this work.

Abstract: Emotions play a very important role in how we think and behave. As such, the emotions we feel every day can compel us to act and influence the decisions and plans we make about our lives. Being able to measure, analyze, and better comprehend how or why our emotions may change is thus of much relevance to understand human behavior and its consequences. Despite the great efforts made in the past in the study of human emotions, it is only now with the advent of wearable, mobile, and ubiquitous technologies that we can aim at sensing and recognizing emotions, continuously and in the wild. This Special Issue aims at bringing together the latest experiences, findings, and developments regarding ubiquitous sensing, modeling, and recognition of human emotions.

Keywords: affective computing; emotion recognition; artificial intelligence; machine learning; image processing; video processing

check for updates

Citation: Banos, O.; Castro, L.A.; Villalonga, C. Ubiquitous Technologies for Emotion Recognition. *Appl. Sci.* **2021**, *1*, 0. https://doi.org/

Received: 21 July 2021
Accepted: 27 July 2021
Published: 29 July 2021

There is an increased interest in studying facial expressions and how they relate to emotions. Lately, some research has been carried out to enhance in-vehicle interactions for drivers. This increased interest has been mainly focused at increasing road safety such as driver drowsiness, distracted driving, or detecting drivers' stress. In [1], they proposed a novel contactless method for detecting stress in drivers. This method is based on the extraction of features from thermal infrared imaging, which were obtained in a controlled environment with 10 subjects. In their work, they compared their results with a stress index obtained from electrocardiography (ECG). They used a binary classifier to discriminate among drivers with stress vs. those with no stress, resulting in classification performance with an AUC of 0.80, a sensitivity of 77%, and a specificity of 78%. The results are promising towards identifying driver states that can derive in situations that can be of interest for road safety. In addition to this work, the authors of [2] implemented a system to monitor the driver's facial expressions to assess health risks such as severe pain episodes while driving. The proposed approach is based on line-segment feature analysis-convolutional recurrent neural network (LFA-CRNN), which had the highest accuracy at approximately 98.92%. The results of this work aim at alerting the driver to risk-related matters while driving.

Furthermore, using facial expressions, Ref. [3] present a novel system for automatically detecting pain in babies using image analysis. They implemented a classifier based on Support Vector Machines (SVM) using the following texture descriptors for pain detection: Local Binary Patterns, Local Ternary Patterns, and Radon Barcodes. Using the Infant COPE database, the authors obtained accuracy around 95%. The approaches of [2,3] both use facial expressions to detect pain, although they use different databases and features. Still, these works provide valuable results that can be applied in systems for increasing and monitoring safety and health.

Image-based emotion recognition is mostly based on the analysis of whole-face images. However, some approaches put the focus on core facial areas (eyes, nose, mouth, etc.), which normally encode most of the affective information. In [4], the authors multi-block deep learning techniques which are trained on core facial areas. They show its application in an app for interview self-management. The same principle is used in [5]. Here, the authors propose the use of convolutional neural networks for the automatic recognition of micro-expressions in real-time. Although the accuracy results are far from optimal, they appear quite promising, especially concerning the time taken to recognize the emotion (a few milliseconds).

Regular RGB cameras are most frequently used in facial expression-based emotion recognition. However, there is a growing bulk of research devoted to using thermal infrared cameras to this very end. In [6], the authors review the application of thermal infrared imaging to determine affective responses ascribed to physiological modulations. Their review outlines the main advantages and challenges of thermal imaging-based affective computing, with a particular focus on its use for human–robot interaction applications.

The automatic recognition of emotions has been largely based, according to the literature, on image or video data. Nonetheless, other types of data are used to identify people's affective states. For example, Ref. [7] analyzes the audio data generated during phone calls in a call center. The authors use speech emotion recognition to determine the call urgency, giving greater priority to calls featuring emotions such as fear, anger and sadness, and less priority to calls featuring neutral speech and happiness. The results show a significant reduction in waiting time for calls estimated as more urgent, especially those calls presenting emotions of fear and anger.

Electroencephalogram (EEG) signals are another alternate data used for emotion recognition. EEG-based emotion recognition is particularly considered due to its effectiveness compared to body expressions and other physiological signals. In [8], the authors study the use of various statistical techniques, namely logistic regression, Gaussian kernels and Laplacian priors to automatically recognize a set of emotions. They also investigate the critical frequency bands in emotion recognition, concluding a superiority of gamma and beta bands while classifying emotions. The great volume of data generated by EEG approaches makes it particularly interesting to consider using deep learning techniques. In [9], the authors precisely use such techniques with the aim to help determining the attitude of consumers towards a product. The authors prototype their solution with a well-known emotion recognition dataset and they show accuracy results comparable to those obtained via traditional machine learning approaches.

Author Contributions: All authors contributed equally to the screening, review, and assessment of the submitted papers. All authors have read and agreed to the published version of the manuscript.

Funding: This research received no external funding.

Institutional Review Board Statement: Not applicable.

Informed Consent Statement: Not applicable.

Data Availability Statement: Not applicable.

Acknowledgments: Thanks to all the authors who submitted their research for this Special Issue. The invaluable contribution of the international reviewers is gratefully acknowledged.

Conflicts of Interest: The authors declare no conflict of interest.

References

1. Cardone, D.; Perpetuini, D.; Filippini, C.; Spadolini, E.; Mancini, L.; Chiarelli, A.M.; Merla, A. Driver Stress State Evaluation by Means of Thermal Imaging: A Supervised Machine Learning Approach Based on ECG Signal. *Appl. Sci.* **2020**, *10*, 5673, doi:10.3390/app10165673.
2. Kim, C.M.; Hong, E.J.; Chung, K.; Park, R.C. Driver Facial Expression Analysis Using LFA-CRNN-Based Feature Extraction for Health-Risk Decisions. *Appl. Sci.* **2020**, *10*, 2956, doi:10.3390/app10082956.
3. Martínez, A.; Pujol, F.A.; Mora, H. Application of Texture Descriptors to Facial Emotion Recognition in Infants. *Appl. Sci.* **2020**, *10*, 1115, doi:10.3390/app10031115.
4. Shin, D.H.; Chung, K.; Park, R.C. Detection of Emotion Using Multi-Block Deep Learning in a Self-Management Interview App. *Appl. Sci.* **2019**, *9*, 4830, doi:10.3390/app9224830.

5. Belaiche, R.; Liu, Y.; Migniot, C.; Ginhac, D.; Yang, F. Cost-Effective CNNs for Real-Time Micro-Expression Recognition. *Appl. Sci.* **2020**, *10*, 4959, doi:10.3390/app10144959.
6. Filippini, C.; Perpetuini, D.; Cardone, D.; Chiarelli, A.M.; Merla, A. Thermal Infrared Imaging-Based Affective Computing and Its Application to Facilitate Human Robot Interaction: A Review. *Appl. Sci.* **2020**, *10*, 2924, doi:10.3390/app10082924.
7. Bojanić, M.; Delić, V.; Karpov, A. Call Redistribution for a Call Center Based on Speech Emotion Recognition. *Appl. Sci.* **2020**, *10*, 4653, doi:10.3390/app10134653.
8. Pan, C.; Shi, C.; Mu, H.; Li, J.; Gao, X. EEG-Based Emotion Recognition Using Logistic Regression with Gaussian Kernel and Laplacian Prior and Investigation of Critical Frequency Bands. *Appl. Sci.* **2020**, *10*, 1619, doi:10.3390/app10051619.
9. Aldayel, M.; Ykhlef, M.; Al-Nafjan, A. Deep Learning for EEG-Based Preference Classification in Neuromarketing. *Appl. Sci.* **2020**, *10*, 1525, doi:10.3390/app10041525.

applied sciences

Article

Driver Stress State Evaluation by Means of Thermal Imaging: A Supervised Machine Learning Approach Based on ECG Signal

Daniela Cardone [1,*], David Perpetuini [1], Chiara Filippini [1], Edoardo Spadolini [2], Lorenza Mancini [2], Antonio Maria Chiarelli [1] and Arcangelo Merla [1,2]

1 Department of Neurosciences, Imaging and Clinical Sciences (DNISC), University G. d'Annunzio of Chieti-Pescara, 66100 Chieti, Italy; david.perpetuini@unich.it (D.P.); chiara.filippini@unich.it (C.F.); antonio.chiarelli@unich.it (A.M.C.); arcangelo.merla@unich.it (A.M.)

2 Next2U s.r.l., 65127 Pescara, Italy; e.spadolini@next2u-solutions.com (E.S.); l.mancini@next2u-solutions.com (L.M.)

* Correspondence: d.cardone@unich.it; Tel.: +39-0871-3556954

Received: 14 July 2020; Accepted: 13 August 2020; Published: 15 August 2020

Featured Application: A procedure for a driver's stress state monitoring was provided by means of thermal infrared imaging. It was validated on ECG-derived parameters through the application of supervised machine learning techniques.

Abstract: Traffic accidents determine a large number of injuries, sometimes fatal, every year. Among other factors affecting a driver's performance, an important role is played by stress which can decrease decision-making capabilities and situational awareness. In this perspective, it would be beneficial to develop a non-invasive driver stress monitoring system able to recognize the driver's altered state. In this study, a contactless procedure for drivers' stress state assessment by means of thermal infrared imaging was investigated. Thermal imaging was acquired during an experiment on a driving simulator, and thermal features of stress were investigated with comparison to a gold-standard metric (i.e., the stress index, SI) extracted from contact electrocardiography (ECG). A data-driven multivariate machine learning approach based on a non-linear support vector regression (SVR) was employed to estimate the SI through thermal features extracted from facial regions of interest (i.e., nose tip, nostrils, glabella). The predicted SI showed a good correlation with the real SI ($r = 0.61$, $p = \sim 0$). A two-level classification of the stress state (STRESS, SI $\geq$ 150, versus NO STRESS, SI < 150) was then performed based on the predicted SI. The ROC analysis showed a good classification performance with an AUC of 0.80, a sensitivity of 77%, and a specificity of 78%.

Keywords: driver stress state; IR imaging; machine learning; support vector machine (SVR); advanced driver-assistance systems (ADAS)

1. Introduction

According to the latest estimates by the World Health Organization, approximately 1.35 million people die each year from road traffic accidents and between 20–50 million people suffer from non-fatal injuries [1].

Advanced driver-assistance systems (ADAS) are designed to support humans during the driving process, leading to an increase in road safety. Conventional ADAS technologies are mainly based on controlling the vehicle state through proprioceptive (i.e., Odometry, inertial sensors) and exteroceptive sensors (i.e., Lidar, vision sensors, radar, infrared, and ultrasonic sensors) [2]. These state-of-the-art technologies allow for the recognition of objects [3], alerting the driver about dangerous road conditions [4], providing driver tips to improve their driving comfort and safety [5], recognizing traffic activity and behavior [6], and detecting risky driving conditions [7].

In addition to factors currently evaluated by ADAS technologies, it is also of fundamental importance to monitor the driver's psycho-physiological state that is strictly related to driving performance as reported by National Highway Traffic Safety Administration (NHTSA) [8]. According to the latest estimation by the NHTSA [8], in the USA, more than 2800 people died and approximately 400,000 people were injured in crashes induced by distracted driving in 2018. In particular, driver drowsiness/fatigue and emotion (i.e., visible anger, sadness, crying, and/or emotional agitation) can increase the risk of car accidents by 3.4 and 9.8 folds, respectively [9].

Driver state monitoring is mainly based on two categories of approaches that depend on the nature of the data collected [10]. The first approach, named the behavioral method, is based on monitoring a driver's parameters including gaze direction, blinking frequency, percentage of eye closure (PERCLOS), yawning, and head pose. These parameters are evaluated by means of one or multiple visible cameras. This procedure, based on cameras, is indeed contactless and non-invasive, but it is characterized by relevant technical challenges derived from occlusion, illumination variation, and personal privacy issues. Nonetheless, because of its utility, it is often employed in the automobile industry. The second approach, labeled the physiological method, is instead based on monitoring driver's vital signals, such as those derived from electrocardiography (ECG) [11], photoplethysmography (PPG) [12], electrooculography (EOG) [13], electroencephalography (EEG) [14], galvanic skin response (GSR) [15], and electromyography (EMG) [12].

With focus on driver stress, it is known that a high stress state decreases decision-making capabilities and situational awareness, impairing driving performance [16]. Electroencephalography has been widely employed to monitor drivers' stress. In particular, human stress can be inferred by measuring an increase in heart rate as well as by measuring variations of parameters associated to heart rate variability (HRV) which, in turn, can dynamically reflect the accumulation of mental workload [17]. Among the variety of indices derived from ECG signals, Baevsky [18] proposed the Stress Index (SI) which is indicative of both sympathetic activity and central regulation.

Although the SI is a sensitive and specific metric to stress, it is based on contact-based technology (e.g., ECG). Indeed, to ensure comfort and non-intrusiveness for drivers, the use of contact sensors for data collection should preferably be avoided. Infrared (IR) or thermal imaging is a passive technology able to evaluate the spontaneous emission of body thermal energy and measure the temperature in a contactless manner. Infrared imaging indeed allows to overcome the limitations of contact devices and, importantly, in comparison with visible cameras, is not affected by illumination and can work in a completely dark environment.

Relevantly to the topic of this study, IR imaging allows to infer the peripheral autonomic activity through the modulation of the cutaneous temperature, which is a known expression of the psycho-physiological state of the subject [19,20]. In fact, experienced emotions, including stress or fatigue, can produce changes in skin temperature [21–23]. Particularly, there is great attention in this research field on stress and mental workload monitoring using thermal IR imaging. Puri et al. [24] studied computer users' stress, reporting an increased blood flow in the frontal vessels of the forehead during stressful situations. The thermal metric was shown to be correlated with stress levels in 12 participants performing a Stroop test ($r = 0.9$, excluding one outlier). Pavlidis and colleagues [25] tried to assess the stress level by measuring transient perspiratory responses on the perinasal area using thermal IR imaging. These metrics proved to be a good indicator of stress response, because they were sympathetically driven. The authors applied this approach in the context of surgical training, finding a very high correlation between the GSR (galvanic skin response) and the thermal measurement on the finger ($r = 0.968$) and on the perinasal region ($r = 0.943$). Kang et al. [26] used thermal IR imaging to assess affective training times by monitoring the cognitive load through facial temperature changes. Learning proficiency patterns were based on an alphanumeric task. Significant correlations, ranging from $r = 0.88$ to $r = 0.96$, were found between the nose tip temperature and the response time, accuracy, and the Modified Cooper Harper Scale ratings. Stemberger et al. [27] presented a system for the estimation of cognitive workload levels based on the analysis of facial skin temperature. Beyond thermal infrared imaging of the face, the system relied on head pose estimation, measurement of the temperature variation across regions of the face, and an artificial neural network classifier. The system was capable of accurately classifying mental workload into high, medium, and low levels 81% of the time.

Given the advantages of the use of thermography in driver state monitoring, a relevant number of scientific works on this research field are available. Most of these publications concern driver drowsiness/fatigue monitoring and emotional state detection. Ebrahimian-Hadikiashari et al. [28] investigated driver drowsiness by analyzing the breathing function, monitored by thermography. The authors observed a significant decrease of driver respiration rate from extreme drowsiness to wakefulness conditions. Moreover, Knapik et al. [29] presented an approach for the evaluation of driver's fatigue, based on yawn detection using thermal imaging. Zhang et al. [30] demonstrated the feasibility of discriminating emotions (e.g., fear versus no fear) by means of thermal imaging, assessing the forehead temperature as indicative for the emotional dimension of drivers' fear. Focusing on driver stress monitoring, Yamakoshi et al. [31] combined measures from facial skin temperature and hemodynamic variables. The authors observed an increase of sympathetic activity, peripheral vasoconstriction, hence, a significant decrease in peripheral skin temperature during monotonous driving simulation. Basing on differential skin temperatures between peripheral (i.e., nose tip) and truncal parts of the face (i.e., cheeks, jaw, and forehead), they were able to assess an index of a driver's stress. More recently, Pavlidis et al. [32] studied the effects of cognitive, emotional, sensorimotor, and mixed stressors on driver arousal and performance with respect to a baseline of 59 drivers in a simulation experiment. Perinasal perspiration, revealed by thermal imaging, together with the measure of steering angle and the range of lane departures on the left and right side of the road, showed a more dangerous driving condition in the case of sensorimotor and mixed stressors with respect to the baseline condition.

In this paper, the driver stress state was established by means of IR imaging and supervised machine learning methods. Supervised machine learning approaches are part of artificial intelligence (AI) algorithms, able to automatically learn functions that map an input to an output based on known input–output pairs (training dataset). The function is inferred from labeled training data and can be used for mapping new dataset (test dataset) that allow to evaluate the accuracy of the learned function and understand the level of generalization of the applied model [33].

On the basis of key features of thermal signals extracted from peculiar regions of interest (ROIs), indicative of the psycho-physiological state, an estimation of the ECG-derived SI was performed employing a support vector regression with radial basis function (SVR-RBF) [17]. To test the generalization performances of the model, a leave-one-subject-out cross-validation was utilized. After the cross-validation process, a two-level classification of the stress state (STRESS versus NO STRESS) was performed, relying on the estimated SI.

This work describes a novel approach for a contactless methodology dedicated to driver stress state detection and classification, constituting a significant improvement to actual ADAS technology and, in general, to road security level.

2. Materials and Methods

2.1. Participants

The experimental session involved 10 adults (6 males, age range 22–35, mean 28.4). Before the start of the experimental trials, the participants were adequately informed about the purpose and protocol of the study, and they signed an informed consent form outlining the methods and the purposes of the experimentation in accordance with the Declaration of Helsinki [34].

2.2. Procedure and Data Acquisition

Prior to testing, each subject was left in the experimental room for 20 min to allow the baseline skin temperature to stabilize. The recording room was set at a standardized temperature (23 °C) and humidity (50–60%) by a thermostat.

To perform the experiment, a static driver simulator was used (Figure 1a). It was composed of driver's seat, steering wheel, clutch, brake, and gas pedals, and gearshift. To display the scenario, three 27 inch monitors were used. The total video resolution for the stimulation was 5760 × 1080 pixels. The distance between the driver and road screen was approximately 1.5 m. Participants' horizontal view angle was 150 degrees. The simulator could produce starter and engine sounds, left and right signal indicators, and flashers and wiper blades. In this study, the sound of the engine, starter, and lights switches were provided.

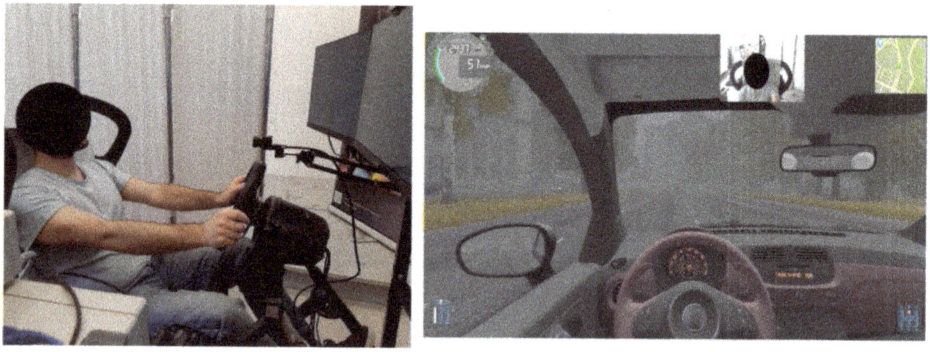

(a) (b)

Figure 1. Experimental setting for the proposed study: (**a**) driving simulator, lateral view; (**b**) screenshot of the software used for the driving simulation: City Car Driving [34].

Participants sat comfortably on the seat of the driving simulator during both acclimatization and measurement periods.

The software used for driving simulation was City Car Driving, Home Edition software (version 1.5) [35] (Figure 1b). The experimental protocol consisted of performing a driving simulation lasting 45 min in an urban context. The experimental conditions were set a priori to ensure the adverse driving condition and guarantee the uniformity of the experimental protocol for all the subjects. An overview of the experimental setting of City Car Driving software is reported in Table 1.

Table 1. Experimental settings of the driving software City Car Driving.

City Car Driving Settings	Conditions
Weather	• Season: Autumn • Weather condition: Foggy • Time of the day: Daytime
Traffic	• Traffic density: 50% • Traffic behavior: Intense traffic • Fullness of traffic: 60%
Territory	• Area: New city • Location: Modern district
Emergency situations	• Dangerous change of traffic: Often • Emergency braking of the car ahead: Often • Pedestrian crossing the road in a wrong place: Often • Accident on the road: Often • Dangerous entrance of the vehicle into the oncoming lane: Rarely

The conditions reported in Table 1 were selected to induce stress in the participants. In particular, the settings associated to traffic and emergency situations guaranteed non-comfortable driving, since the participants were often experiencing non-monotonous situations.

During the execution of the experimental protocol, ECG signals and visible and thermal IR videos were acquired.

The ECG signals were recorded by means of AD Instruments PowerLab system using the lead configuration determined by the Standard Limb Leads (i.e., electrodes positioned at the right arm (RA), left arm (LA), and left leg (LL)) [36].

Visible and thermal IR videos were acquired by the depth camera Intel RealSense D415 and FLIR Boson 320LW IR thermal camera, respectively. The technical characteristics of the two acquisition devices are summarized in Table 2.

Table 2. Technical characteristics of the depth camera Intel RealSense D415 and FLIR Boson 320LW IR thermal camera.

Technical Data	Intel RealSense D415	FLIR Boson 320 LWIR
Weight	4.54 g	7.5 g without lens
Dimensions	99 × 20 × 23 mm	21 × 21 × 11 mm without lens
Spatial resolution	Full HD 1080p (1920 × 1080)	320 × 256
Acquisition rate	30 fps @ 1080p	30 fps
Field of view (FOV)	69.4° × 42.5° × 77° (±3°)	92° HFoV [1]
Sensors technology	Rolling Shutter, 1.4 μm × 1.4 μm pixel size	Uncooled VOx microbolometer
Thermal Sensitivity	-	<50 mK (Professional)

[1] Horizontal field of view.

For the purpose of this study, the two imaging devices were held together and aligned horizontally. Figure 2 shows the entire imaging acquisition system.

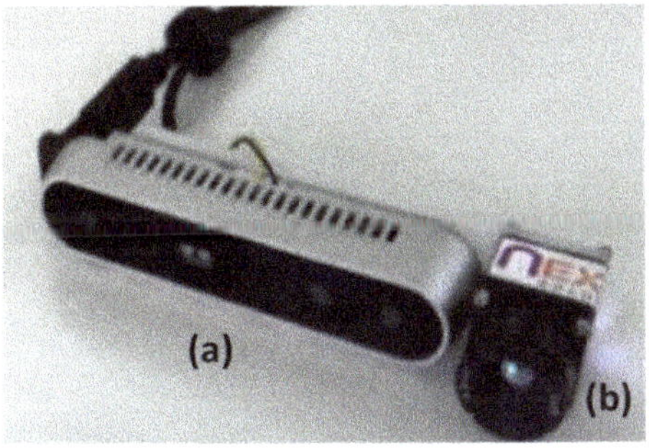

Figure 2. Imaging acquisition system: (**a**) depth visible camera and (**b**) thermal camera.

2.3. *Analysis of ECG Signals*

The ECG signals were recorded at a rate of 1 kHz. The elapsed time periods between the two successive R-peaks of the ECGs (RR signals) were extracted from LabChart7, ADInstruments, and analyzed by the software Kubios HRV Standard [37]. Baevsky's Stress Index (SI) [18] was evaluated for each subject in 30 s consecutive windows. The SI from Kubios is the square root (to make the index normally distributed) of the Baevsky's Stress Index proposed in Reference [18].

Baevsky's SI is calculated based on the distribution of the RR intervals as reported in Equation (1):

$$SI = \frac{AMo \times 100\%}{2Mo \times MxDMn} \tag{1}$$

where Mo is the mode (the most frequent RR interval), AMo is the mode amplitude expressed in percent, and $MxDMn$ is the variation scope reflecting the degree of RR interval variability.

Values of Baevsky's stress index between 80 and 150 are considered normal [18].

2.4. *Analysis of Visible and Thermal Imaging Data*

Visible and IR videos of the subjects' faces were simultaneously recorded during the driving experiment at an acquisition frame rate of 30 Hz and 10 Hz, respectively.

Given the availability of computer vision algorithms for visible videos, in the present study, visible imaging was used as reference for tracking facial landmark features. The purpose of the visible tracking was to transfer the visible facial landmark features tracked to the thermal imagery, estimating the geometrical transformation between the two imaging optical devices.

2.4.1. Visible and Thermal Data Co-Registration

The first step of the developed procedure consisted of an optical co-registration between visible and thermal optics. The co-registration process was a fundamental step of the whole pipeline, since it allowed a proper mapping from an imaging coordinate system to another.

The optical co-registration relied on procedures implemented in OpenCV [38], and it is described in depth in Reference [39]. A root mean square error (RMSE) value was provided by the co-registration procedure, thus indicating the accuracy in the coordinate transformation from visible to IR imagery at the specific distance of 1 m.

2.4.2. Facial Landmark Detection in the Visible Domain

Visible videos were analyzed through OpenFace [40,41], an open-source software able to perform facial landmark detection, head pose estimation, facial action unit recognition, and eye-gaze estimation. For each frame, a set of 68 facial landmarks was estimated during the experiment. Figure 3 shows the distribution of the 68 facial landmarks.

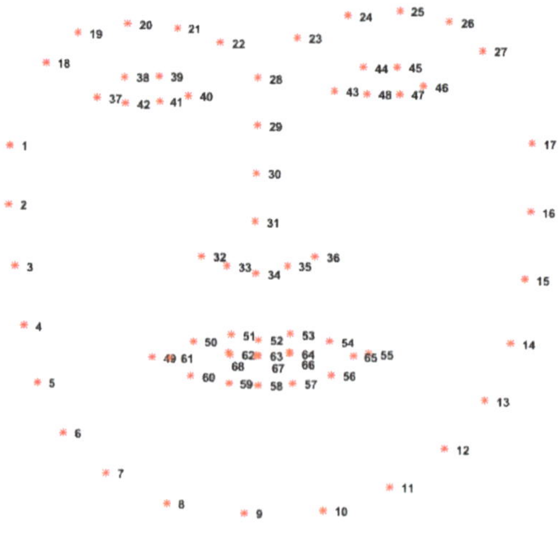

Figure 3. Schematic representation of the 68 facial landmarks identified by the algorithm implemented in the OpenFace software.

The landmark detector algorithm within OpenFace relied on the constrained local neural fields (CLNF) procedure [42], whereas the face detector algorithm employed a multi-task convolutional cascaded network (MTCNN) approach [43].

2.4.3. Thermal Data Extraction and Analysis

The sets of the 68 facial landmarks detected in the visible images were identified in the corresponding frames of IR imaging, applying the geometrical transformation obtained from the optical co-registration process. Figure 4a,b show an example of the 68 feature landmarks detected on a visible frame and the set of the 68 points identified on the corresponding thermal image.

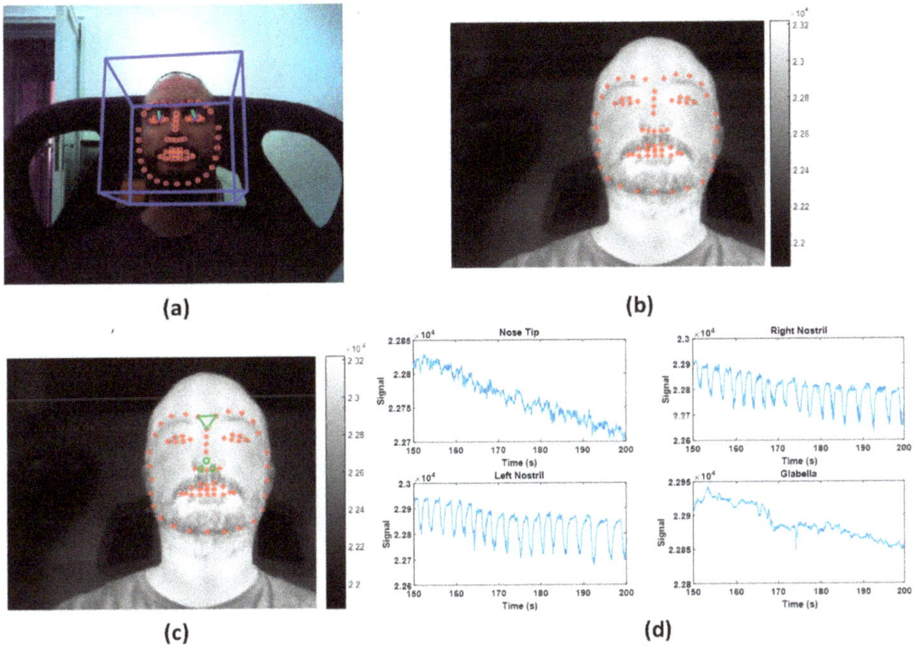

Figure 4. (**a**) Facial landmark identification in the visible image by OpenFace; (**b**) facial landmark identification in the corresponding thermal image applying the geometrical transformation obtained from the optical co-registration process; (**c**) Regions Of Interest (ROIs) identification (nose tip, right and left nostrils, glabella); (**d**) average thermal signals extracted from the ROIs in an exemplificative time window of 50 s. Notice the breathing signal is clearly appreciable from the right and left nostrils' average thermal signals plots.

A fundamental aspect for obtaining an accurate co-registration was the need of temporal synchronization between visible and IR videos. Since the acquisition frame rate of thermal videos was lower than that of the visible camera, the corresponding frames within the visible domain were determined according to the specific timestamps of IR frames. Specifically, among the visible frames acquired around the IR frame timestamp, the one that minimized the temporal difference with IR imaging was chosen. The timestamps of the frames were considered reliable as the videos were acquired on the same PC.

For each thermal video, four ROIs were considered and positioned on facial areas of physiological importance (nose tip, right and left nostrils, and glabella) [44]. The ROIs' coordinates were automatically determined from the location of the 68 landmarks. In this way, the initialization of the position of the ROIs was automatically determined (Figure 4c).

With reference to the topographical distribution of the points as represented in Figure 3, the coordinates of the four ROIs were determined, as described in Table 3:

Table 3. Geometrical features of the considered ROIs.

Region of Interest (ROI)	ROI Shape	ROI Position Relative to 68 Facial Landmark
ROI 1—Nose tip	Circle	$C = \left[\frac{x_{31}+x_{34}}{2}, \frac{y_{31}+y_{34}}{2}\right]$, d = 7 pixel [1]
ROI 2—Right nostril	Circle	$C = [x_{33}, y_{33}]$, d = 7 pixel [1]
ROI 3—Left nostril	Circle	$C = [x_{35}, y_{35}]$, d = 7 pixel [1]
ROI 4—Glabella	Polygon	Polyline ([P22, P23, P28]) [2]

[1] C = circle center; d = circle diameter; [2] Pn = n-th landmark; $n = 1, \ldots, 68$.

For each ROI, the average value of the pixels was extracted over time (Figure 4d). Relatively to the nostrils' ROIs, the average value between ROI 2 and ROI 3 was considered for further statistical analysis, them being related to the same physiological process (i.e., breathing function).

For each of the extracted signals, six representative features were computed over consecutive temporal window of 30 s:

(1) Absolute value of the difference between the average of the signal in the first 5 s and in the last 5 s (Δ);
(2) Standard deviation of the raw thermal signals (STD);
(3) The 90th percentile of the raw thermal signals (90th P);
(4) Kurtosis of the raw thermal signals (K);
(5) Skewness of the raw thermal signals (S);
(6) Ratio of the power spectral density of the raw thermal signals evaluated in the low-frequency band (LF = (0.04–0.15) Hz) and in the high-frequency band (HF = (0.15–0.4) Hz) (LF/HF).

2.4.4. Application of Supervised Machine Learning

Firstly, a machine learning approach was utilized to predict SI relying on features extracted from thermal signals. Specifically, an SVR with RBF kernel was trained on the SI obtained from Kubios through a supervised learning procedure. The SVR-RBF was trained on z-scored data with a fixed nonlinearity exponential parameter $\gamma = 1$.

Because of the multivariate (6 regressors) SVR approach, in-sample performance of the procedure did not reliably estimate the out-of-sample performance. The generalization capabilities of the procedure were thus assessed through cross-validation. Specifically, a leave-one-subject-out cross-validation was performed [45]. This cross-validation procedure consisted in leaving one subject (specifically all the samples from the same subject) out of the regression and in estimating the predicted output value on the given subject using the other participants as the training set of the SVR model. This procedure was iterated for all the subjects, and further statistical analyses were performed on the out-of-training-sample estimation of SI from thermal features. Such a metric was labelled SIcross.

Although several machine learning approaches could be suited for such a purpose, given the limited number of independent features available and the exploratory nature of the implemented approach, an SVR-RBF followed by a classification procedure was chosen to limit the procedural complexity. In fact, although SVR-RBF is not a sophisticated approach, it ensures performances which are comparable to more complex machine learning techniques [46].

Secondly, SIcross was used to perform a two-level classification of the driver's stress (i.e., STRESS versus NO STRESS). The two classes were defined on the basis of the threshold associated to a stress condition assessed by the SI (i.e., SI > 150 for stress condition) [18]. Notably, the experimental recordings confirmed the accordance between SI and the driving conditions. In particular, stressful situations assessed by SI were associated to adverse events during driving simulations (e.g., traffic accidents, collisions with pedestrians, sudden car braking).

Since the two classes did not have an equal number of samples, a bootstrap procedure was implemented to test classification performance on balanced classes [47]. The performances of the classification were evaluated by means of receiver operating characteristic (ROC) analysis [48].

Figure 5 reports the flow chart relating to the described machine learning approach.

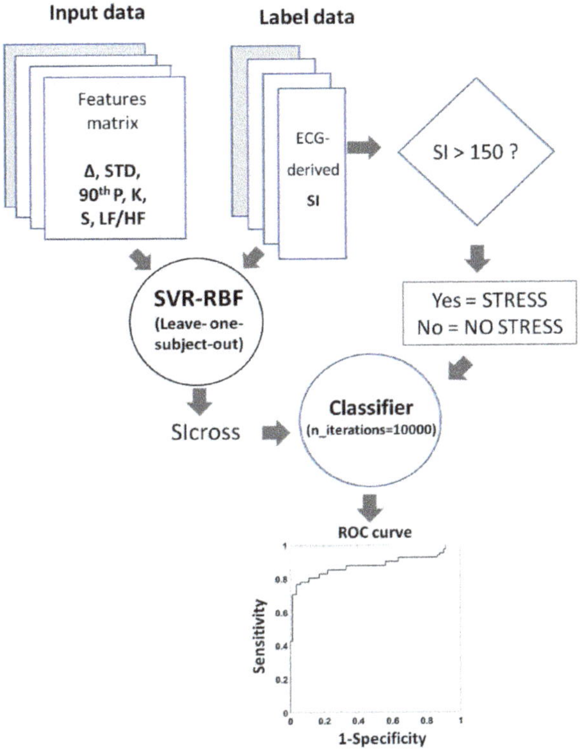

Figure 5. Flow chart of the applied machine learning approach: the thermal features are used as predictors whilst the Electrocardiogram(ECG)-derived Stress Index (SI) is considered as the regression output. Support Vector Regression with Radial Basis Function kernel (SVR-RBF) was used as regressor. A leave-one-subject-out cross-validation was employed to test the generalization of the regression. The result of the regression (SIcross) was then used to perform a two-level classification of the driver's stress (i.e., STRESS versus NO STRESS). Since the two classes were not balanced, a bootstrap procedure was implemented. Receiver Operating Characteristic (ROC) analysis was executed to investigate the performance of the classifier.

3. Results

3.1. Visible and Thermal Imaging Co-Registration and Processing

The spatial RMSE of the optical co-registration was 0.66 ± 0.25 pixels, thus indicating that the accuracy in the coordinate transformation from visible to IR imagery at the specific distance of one meter was less than one pixel.

The percentage value of the correctly identified landmark on the total amount of considered frames and the confidence value in correctly classifying a face are reported in Table 4 for each subject. These parameters were returned by the software OpenFace [40]. The confidence value ranged from 0 (total misclassification of face) to 1 (correct face classification), and it was the result of a landmark detection validation process. In detail, to avoid tracking drift over time, it was necessary to determine if landmark detection succeeded during video processing. The landmark detection validation was performed transforming the area surrounded by the landmarks to a pre-defined reference shape. The vectorized resulting image was then used as a feature vector for a classifier which acts as the validator (i.e., input of the classifier). To train the classifier on the vectorized reference warp, positive and negative landmark detection examples were considered. The positive samples were ground truth landmark labels, whereas the negative samples were generated from the ground truth labels, applying offset and scale transformations. The classifier employed in OpenFace is SVM [49].

Table 4. Indices of performance of landmark identification and face classification on visible imaging.

Subject ID	Success (%)	Confidence
Subject 01	100.00	0.93
Subject 02	99.87	0.98
Subject 03	77.90	0.76
Subject 04	99.97	0.98
Subject 05	70.54	0.66
Subject 06	98.94	0.96
Subject 07	99.87	0.91
Subject 08	99.80	0.93
Subject 09	99.86	0.97
Subject 10	99.81	0.96

On average, 94.66% of the video frames were correctly processed, whereas the confidence index for face classification was 0.90%.

To notice, concerning subjects 3 and 5, the average success and confidence scores were lower with respect to the other subjects, given the scarce lighting conditions of the acquisitions (Table 4). However, in general, for all the subjects, only the frames with high confidence and success scores were considered for further analysis (i.e., success index > 90%, confidence value > 0.8). This ensures there was no impact on the ROIs' identification and, consequently, on their features' estimation.

Finally, the average execution time of the developed algorithm was 0.09 s/frames with MATLAB 2016b© (64-bit Windows 7 Pro, Service Pack 1; Intel (R) Core (TM) i5 CPU; 8.00 GB RAM).

3.2. *Performances of Supervised Machine Learning Approach*

Across subjects, 849 samples were available for the regression analysis. A significant correlation between SI and predicted SI (SIcross) was obtained ($r = 0.61$, $p = \sim 0$) (Figure 6a), demonstrating a good performance of the multivariate analysis [50]. The weights associated to each z-scored regressor for each ROI are shown in Figure 6b. Considering that both the regressors and SI were normalized, the values of the weights were indicative of the contribution of each model input in the estimation of the SI.

Since the two classes had a different number of samples (125 samples for STRESS versus 696 samples for NO STRESS conditions), a bootstrap procedure was implemented to provide a classification estimates using balanced classes [47].

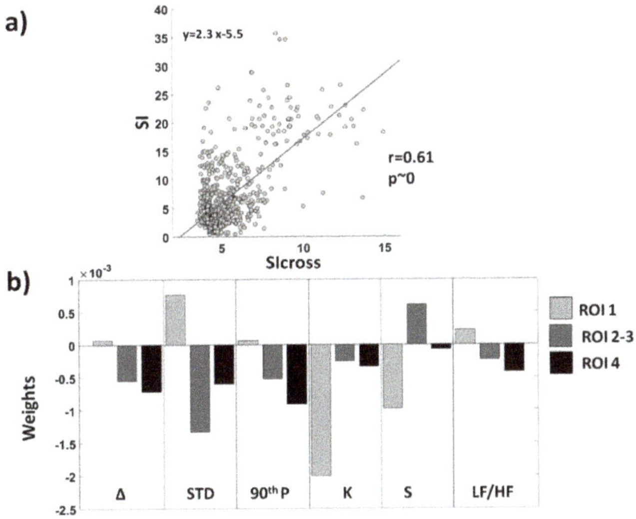

Figure 6. (**a**) Correlation plot between SI and SIcross. The equation of the interpolating line is reported in the top left section of the graph. A good performance of the multivariate analysis is revealed by the correlation score ($r = 0.61$, $p = \sim 0$); (**b**) weights associated to each z-scored regressor for each ROI. The weights are indicative of the contribution of each model input in the estimation of the SI.

Figure 7a reports the among iterations average ROC curve (bootstrap performed for $n = 10{,}000$ iterations). The average area under curve (AUC) was 0.80 and with standard deviation of 0.01. The distribution of the AUC obtained after the bootstrap is reported in Figure 7b.

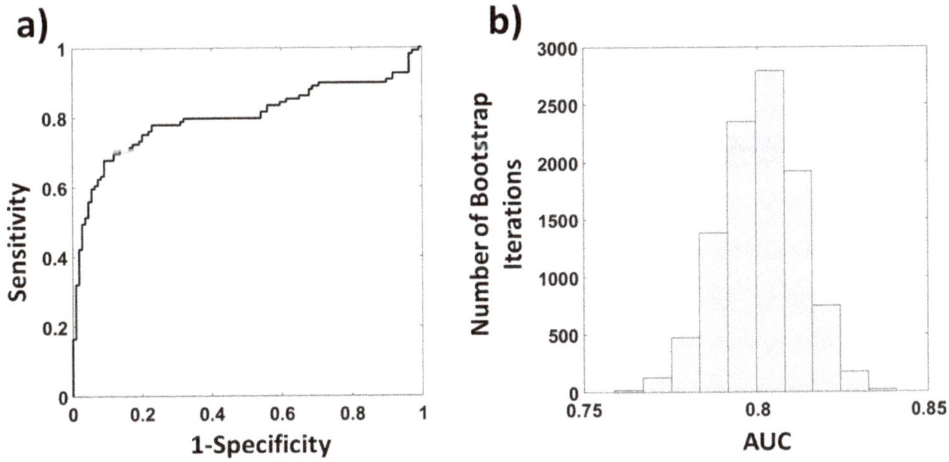

Figure 7. Results after bootstrap procedure (n_iterations = 10,000). (**a**) Among iteration average ROC curve and (**b**) distribution of the Area Under Curve (AUC) obtained after the bootstrap procedure. The average AUC was 0.80 with a standard deviation of 0.01.

By choosing a specific threshold for SIcross, a sensitivity of 77% and a specificity of 78% were obtained as reported in the confusion matrix (Table 5).

Table 5. Confusion matrix of the classification procedure.

Conditions	NO STRESS	STRESS
NO STRESS	78%	22%
STRESS	23%	77%

4. Discussion

In this study, a novel method for driver stress evaluation based on thermal IR imaging and supervised machine learning approaches was described. Thermal IR imaging and ECG were acquired on ten subjects, while performing an experiment on a driving simulator using the software City Car Driving v.1.5 [35]. The experimental session consisted of 45 min of urban context driving with pre-established weather and traffic conditions. Electrocardiography (ECG) signals were used to infer the stress condition of the drivers. Among the variety of indices derived from the ECG signals, stress was considered [18]. In this study, the SI was evaluated in consecutive 30 s time windows by the software Kubios [36]. In the same temporal window, six representative features from average thermal signals on four ROIs (i.e., nose tip, left and right nostrils, glabella) were extracted. The thermal signals were automatically determined by a real-time tracking procedure. The tracking relied on state-of-the-art computer vision algorithms applied on visible images and the optical co-registration between the visible and thermal imaging devices ensuring high performance on signal processing and speed of extraction. Indeed, the high performances were highlighted by the percentage of correctly processed frames which reached an average of 94.66% by the confidence index for face classification that was 0.90 and by the average processing time that was only 0.09 s/frames.

A multivariate machine learning approach based on Support Vector Regression (SVR) with Radial Basis Function (RBF) kernel was employed to estimate the ECG-based SI through peculiar thermal features extracted from facial ROIs. Those ROIs were chosen on the basis of their physiological importance for stress detection [43]. A total amount of 18 thermal features (six features for each ROIs) were computed and used as predictors, while the SI, evaluated through the ECG signals, was considered as the regression output. A leave-one-subject-out cross-validation was employed to test the generalization of the regression. This procedure was iterated for all the subjects and further statistical analyses were performed on the out-of-training-sample estimation of SI. Such a metric was labeled as SIcross. The correlation between SI and SIcross was $r = 0.61$ ($p = {\sim}0$) thus indicating a good estimation of the SI through the considered thermal features (Figure 6a). A feature-based analysis was performed to investigate the relevance of each feature (Figure 6b).

Concerning the nose tip region, the most contributing features to the SI estimation were the kurtosis, the skewness and the standard deviation. The weights associated to the kurtosis and skewness had negative values, thus indicating an inverse relation between SI and the features' trends. The opposite trend was observed for the weights associated to the standard deviation. This pattern seems to be correlated with sweating or vasoconstriction phenomena, occurring with increasing stress [51,52]. In fact, an increase in the standard deviation and a reduction of the kurtosis and skewness parameters (i.e., flatness and asymmetry of the distribution of the related signal [53]) can be associated to a decrease of uniformity of the signal, thus indicating the presence, for instance, of "cold spots" typically present during sweating and vasoconstriction processes [19]. Concerning the nostrils region, instead, a strong inverse relation between the weight associated to the standard deviation and the SI was found. Since the thermal signals from nostrils are highly related to the breathing function, a lower signal variation (revealed by a decrease in the standard deviation) could be associated to a high breathing rate [54]. This result is in accordance with the findings from References [55,56], where it was shown that stress is associated with an increased respiratory rate. Finally, referring to the glabella region, the most relevant feature to the SI estimation was the 90th percentile, i.e., the value below which 90% of data falls. The weight associated to the 90th percentile was directly related to the SI, thus indicating that an increase in temperature of the glabella could be indicative of a stress condition.

This result is in accordance with the findings reported in Reference [57] in which an increase in forehead temperature was associated to the execution of high difficult tasks.

To be noted, when using non-linear regressors, the contribution of each feature in predicting the output does not only depend on the relative weight but also on the non-linearities of the model. Nonetheless, the SVR-RBF employed a single parameter depicting the non-linearity extent for all the features considered. Thus, although not directly regressing the input with the output, the weights of RBF-SVR were still associated to the importance of each regressor.

The SIcross was, then, used to perform a two-level classification of the driver's stress (i.e., STRESS versus NO STRESS). The two classes were defined on the basis of the threshold associated to a stress condition assessed by Baevsky's SI [18]. Since the two classes were not balanced, a bootstrap procedure was implemented [47]. The ROC analysis showed a good performance of the classifier with an average AUC of 0.80 (Figure 7b), a sensitivity of 77%, and a specificity of 78% (Table 4).

It is worth noting that the cross-validation and the bootstrap procedures provided the generalization performances of the model, testing its applicability to a wide cohort of drivers. In fact, although stress conditions could elicit different physiological responses among subjects, for ADAS applications, it could be more relevant to detect stress conditions across participants, rather than focusing on a single subject's stress level.

The main benefit of the developed method with respect to the available literature is the use of supervised machine learning approaches, based on the only thermal features, without accounting for vehicle- or driver's behavioral-related parameters, reaching performances comparable with more complex approaches [27]. Furthermore, the developed method opens the way to efficient real-time implementation of drivers' stress state monitoring relying only on thermal IR imaging, being the model already validated and ready to use.

Nonetheless, further studies should be performed to increase the sample size. The machine learning approach used in this study relied on supervised learning which is inherently a data-driven analysis; data-driven analysis is highly affected by the sample size and the performance of the model could indeed improve reducing a possible overfitting effect driven by the limited sample numerosity. To be noted, the present study focused on drivers with a limited age range (i.e., 22–35 years old), involving only young subjects. The most important improvement of the method will be to include in the study sample people with a wider age range. In future studies, beyond increasing the sample size and age range, other factors, such as gender, thermal comfort, and weather conditions during simulated driving sessions, will be considered [58–60]. In fact, taking these factors into account could be of fundamental valence in driving stress research, leading to a wide overview of all the aspects concerning the matter of the study.

Furthermore, the present results are relative to simulated driving conditions in which determining variables for IR measurements, like direct ventilation or sunlight, were not considered. Thus, it would be desirable to apply the developed methodology also on real-driving situations, to generalize the applicability of the technique.

As for being state-of-the-art, this is an original and novel study concerning drivers' stress state evaluation by means of thermal imaging, employing supervised machine learning algorithms. This is a preliminary study, addressed to limited and specific experimental conditions which, however, underlines the feasibility of the method to be verified under wider operating situations.

5. Conclusions

In the present work, a novel and original method allowing for drivers' stress state evaluation was presented. By using machine learning approaches, it was possible to understand and classify, with a good level of accuracy, the stress state of the subjects while driving in a simulated environment. The presented work constitutes the first step towards the establishment of a reliable detection of the stress s in a non-invasive fashion, ensuring to maintain an ecologic condition during measurements.

Author Contributions: Conceptualization, D.C., D.P., C.F., A.M.C., A.M.; methodology, D.C., D.P.; software, D.C., D.P., E.S.; validation, D.P.; formal analysis, D.C., D.P.; investigation, D.C., D.P., C.F., A.M.C., L.M.; writing—original draft preparation, D.C., C.F., D.P.; writing—review and editing, A.M.C., A.M.; supervision, A.M.; project administration, A.M.; funding acquisition, A.M. All authors have read and agreed to the published version of the manuscript.

Funding: This research was funded by the grants: PON FESR MIUR R&I 2014-2020-ADAS+, grant number ARS01_00459 and ECSEL Joint Undertaking (JU) European Union's Horizon 2020 Heliaus, grant number 826131.

Conflicts of Interest: The authors declare no conflict of interest.

References

1. WHO. *Global Status Report on Road Safety 2018: Summary (No. WHO/NMH/NVI/18.20)*; WHO: Geneva, Switzerland, 2018.
2. Bengler, K.; Dietmayer, K.; Farber, B.; Maurer, M.; Stiller, C.; Winner, H. Three decades of driver assistance systems: Review and future perspectives. *IEEE Intell. Transp. Syst. Mag.* **2014**, *6*, 6–22. [CrossRef]
3. Weon, I.-S.; Lee, S.-G.; Ryu, J.-K. Object Recognition based interpolation with 3d lidar and vision for autonomous driving of an intelligent vehicle. *IEEE Access* **2020**, *8*, 65599–65608. [CrossRef]
4. Catten, J.C.; McClellan, S. System and Method for Alerting Drivers to Road Conditions. U.S. Patent 8,188,887, 29 May 2012.
5. Damiani, S.; Deregibus, E.; Andreone, L. Driver-vehicle interfaces and interaction: Where are they going? *Eur. Transp. Res. Rev.* **2009**, *1*, 87–96. [CrossRef]
6. Huynh-The, T.; Banos, O.; Le, B.-V.; Bui, D.-M.; Yoon, Y.; Lee, S. Traffic behavior recognition using the pachinko allocation model. *Sensors* **2015**, *15*, 16040–16059. [CrossRef] [PubMed]
7. Cruz, L.C.; Macías, A.; Domitsu, M.; Castro, L.A.; Rodríguez, L.-F. Risky driving detection through urban mobility traces: A preliminary approach. In *Context-Awareness and Context-Driven Interaction, Proceedings of the Ubiquitous Computing and Ambient Intelligence, Carrillo, CR, USA, 2–6 December 2013*; Urzaiz, G., Ochoa, S.F., Bravo, J., Chen, L.L., Oliveira, J., Eds.; Springer: Cham, Switzerland, 2013; pp. 382–385.
8. Distracted Driving. Available online: https://www.nhtsa.gov/risky-driving/distracted-driving (accessed on 14 August 2020).
9. Dingus, T.A.; Guo, F.; Lee, S.; Antin, J.F.; Perez, M.; Buchanan-King, M.; Hankey, J. Driver crash risk factors and prevalence evaluation using naturalistic driving data. *Proc. Natl. Acad. Sci. USA* **2016**, *113*, 2636–2641. [CrossRef]
10. Guettas, A.; Ayad, S.; Kazar, O. Driver state monitoring system: A review. In Proceedings of the 4th International Conference on Big Data and Internet of Things, Tangier, Morocco, 23–24 October 2019; pp. 1–7.
11. Minhad, K.N.; Ali, S.H.M.; Reaz, M.B.I. Happy-anger emotions classifications from electrocardiogram signal for automobile driving safety and awareness. *J. Transp. Health* **2017**, *7*, 75–89. [CrossRef]
12. Lee, B.G.; Chong, T.W.; Lee, B.L.; Park, H.J.; Kim, Y.N.; Kim, B. Wearable mobile-based emotional response-monitoring system for drivers. *IEEE Trans. Hum.-Mach. Syst.* **2017**, *47*, 636–649. [CrossRef]
13. Barua, S.; Ahmed, M.U.; Ahlström, C.; Begum, S. Automatic driver sleepiness detection using EEG, EOG and contextual information. *Expert Syst. Appl.* **2019**, *115*, 121–135. [CrossRef]
14. Zeng, H.; Yang, C.; Dai, G.; Qin, F.; Zhang, J.; Kong, W. EEG classification of driver mental states by deep learning. *Cogn. Neurodyn.* **2018**, *12*, 597–606. [CrossRef]
15. Chen, L.; Zhao, Y.; Ye, P.; Zhang, J.; Zou, J. Detecting driving stress in physiological signals based on multimodal feature analysis and kernel classifiers. *Expert Syst. Appl.* **2017**, *85*, 279–291. [CrossRef]
16. Healey, J.A.; Picard, R.W. Detecting stress during real-world driving tasks using physiological sensors. *IEEE Trans. Intell. Transp. Syst.* **2005**, *6*, 156–166. [CrossRef]
17. Munla, N.; Khalil, M.; Shahin, A.; Mourad, A. Driver stress level detection using HRV analysis. In Proceedings of the 2015 International Conference on Advances in Biomedical Engineering (ICABME), Beirut, Lebanon, 16–18 September 2015; pp. 61–64.
18. Baevsky, R.M.; Chernikova, A.G. Heart rate variability analysis: Physiological foundations and main methods. *Cardiometry* **2017**, 66–76. [CrossRef]

19. Cardone, D.; Merla, A. New frontiers for applications of thermal infrared imaging devices: Computational psychopshysiology in the neurosciences. *Sensors* **2017**, *17*, 1042. [CrossRef] [PubMed]
20. Filippini, C.; Perpetuini, D.; Cardone, D.; Chiarelli, A.M.; Merla, A. Thermal infrared imaging-based affective computing and its application to facilitate human robot interaction: A review. *Appl. Sci.* **2020**, *10*, 2924. [CrossRef]
21. Engert, V.; Merla, A.; Grant, J.A.; Cardone, D.; Tusche, A.; Singer, T. Exploring the use of thermal infrared imaging in human stress research. *PLoS ONE* **2014**, *9*, e90782. [CrossRef]
22. Cruz-Albarran, I.A.; Benitez-Rangel, J.P.; Osornio-Rios, R.A.; Morales-Hernandez, L.A. Human emotions detection based on a smart-thermal system of thermographic images. *Infrared Phys. Technol.* **2017**, *81*, 250–261. [CrossRef]
23. Perpetuini, D.; Cardone, D.; Bucco, R.; Zito, M.; Merla, A. Assessment of the Autonomic Response in Alzheimer's Patients During the Execution of Memory Tasks: A Functional Thermal Imaging Study. Available online: https://www.ingentaconnect.com/content/ben/car/2018/00000015/00000010/art00007 (accessed on 25 June 2020).
24. Puri, C.; Olson, L.; Pavlidis, I.; Levine, J.; Starren, J. StressCam: Non-contact measurement of users' emotional states through thermal imaging. In *CHI'05 Extended Abstracts on Human Factors in Computing Systems*; ACM: Portland, OR, USA, 2005; pp. 1725–1728.
25. Pavlidis, I.; Tsiamyrtzis, P.; Shastri, D.; Wesley, A.; Zhou, Y.; Lindner, P.; Buddharaju, P.; Joseph, R.; Mandapati, A.; Dunkin, B.; et al. Fast by nature-how stress patterns define human experience and performance in dexterous tasks. *Sci. Rep.* **2012**, *2*, 305. [CrossRef]
26. Kang, J.; McGinley, J.A.; McFadyen, G.; Babski-Reeves, K. Determining learning level and effective training times using thermography. In Proceedings of the Army Science Conference, Orlando, FL, USA, 27–30 November 2006.
27. Stemberger, J.; Allison, R.S.; Schnell, T. Thermal imaging as a way to classify cognitive workload. In Proceedings of the 2010 Canadian Conference on Computer and Robot Vision, Ottawa, ON, Canada, 31 May–2 June 2010; pp. 231–238.
28. Ebrahimian-Hadikiashari, S.; Nahvi, A.; Homayounfard, A.; Bakhoda, H. Monitoring the variation in driver respiration rate from wakefulness to drowsiness: A non-intrusive method for drowsiness detection using thermal imaging. *J. Sleep Sci.* **2018**, *3*, 1–9.
29. Knapik, M.; Cyganek, B. Driver's fatigue recognition based on yawn detection in thermal images. *Neurocomputing* **2019**, *338*, 274–292. [CrossRef]
30. Zhang, M.; Ihme, K.; Drewitz, U. Discriminating drivers' emotions through the dimension of power: Evidence from facial infrared thermography and peripheral physiological measurements. *Transp. Res. Part F Traffic Psychol. Behav.* **2019**, *63*, 135–143. [CrossRef]
31. Yamakoshi, T.; Yamakoshi, K.; Tanaka, S.; Nogawa, M.; Park, S.B.; Shibata, M.; Sawada, Y.; Rolfe, P.; Hirose, Y. Feasibility study on driver's stress detection from differential skin temperature measurement. In Proceedings of the 2008 30th Annual International Conference of the IEEE Engineering in Medicine and Biology Society, Vancouver, BC, Canada, 20–24 August 2008; pp. 1076–1079.
32. Pavlidis, I.; Dcosta, M.; Taamneh, S.; Manser, M.; Ferris, T.; Wunderlich, R.; Akleman, E.; Tsiamyrtzis, P. Dissecting driver behaviors under cognitive, emotional, sensorimotor, and mixed stressors. *Sci. Rep.* **2016**, *6*, 25651. [CrossRef] [PubMed]
33. Praveena, M.; Jaiganesh, V. A literature review on supervised machine learning algorithms and boosting process. *Int. J. Comput. Appl.* **2017**, *169*, 32–35. [CrossRef]
34. World Medical Association Declaration of Helsinki: Ethical principles for medical research involving human subjects. *JAMA* **2000**, *284*, 3043–3045. [CrossRef]
35. City Car Driving—Car Driving Simulator, PC Game. Available online: https://citycardriving.com/ (accessed on 26 June 2020).
36. Conover, M.B. *Understanding Electrocardiography*; Elsevier Health Sciences: St. Luis, MO, USA, 2002; ISBN 978-0-323-01905-7.

37. Tarvainen, M.P.; Niskanen, J.-P.; Lipponen, J.A.; Ranta-aho, P.O.; Karjalainen, P.A. Kubios HRV—A Software for Advanced Heart Rate Variability Analysis. In Proceedings of the 4th European Conference of the International Federation for Medical and Biological Engineering, Antwerp, Belgium, 23–27 November 2008; Vander Sloten, J., Verdonck, P., Nyssen, M., Haueisen, J., Eds.; Springer: Berlin/Heidelberg, Germany, 2009; pp. 1022–1025.
38. Bradski, G.; Kaehler, A. *Learning OpenCV: Computer Vision with the OpenCV Library*; O'Reilly Media, Inc.: Champaign, IL, USA, 2008; ISBN 978-0-596-55404-0.
39. Filippini, C.; Spadolini, E.; Cardone, D.; Bianchi, D.; Preziuso, M.; Sciarretta, C.; del Cimmuto, V.; Lisciani, D.; Merla, A. Facilitating the child–robot interaction by endowing the robot with the capability of understanding the child engagement: The case of mio amico robot. *Int. J. Soc. Robot.* **2020**, 1–13. [CrossRef]
40. Baltrušaitis, T.; Robinson, P.; Morency, L.-P. OpenFace: An open source facial behavior analysis toolkit. In Proceedings of the 2016 IEEE Winter Conference on Applications of Computer Vision (WACV), Lake Placid, NY, USA, 7–9 March 2016; pp. 1–10.
41. Amos, B.; Ludwiczuk, B.; Satyanarayanan, M. Openface: A general-purpose face recognition library with mobile applications. *CMU Sch. Comput. Sci.* **2016**, *6*, 1–18.
42. Baltrusaitis, T.; Robinson, P.; Morency, L.-P. Constrained Local Neural Fields for Robust Facial Landmark Detection in the Wild. In Proceedings of the IEEE International Conference on Computer Vision Workshops, Sydney, Australia, 2–8 December 2013; pp. 354–361.
43. Zhang, K.; Zhang, Z.; Li, Z.; Qiao, Y. Joint Face Detection and Alignment Using Multitask Cascaded Convolutional Networks. *IEEE Signal Process. Lett.* **2016**, *23*, 1499–1503. [CrossRef]
44. Ioannou, S.; Gallese, V.; Merla, A. Thermal infrared imaging in psychophysiology: Potentialities and limits. *Psychophysiology* **2014**, *51*, 951–963. [CrossRef]
45. Vehtari, A.; Gelman, A.; Gabry, J. Practical Bayesian model evaluation using leave-one-out cross-validation and WAIC. *Stat. Comput.* **2017**, *27*, 1413–1432. [CrossRef]
46. Crone, S.F.; Guajardo, J.; Weber, R. A study on the ability of support vector regression and neural networks to forecast basic time series patterns. In Proceedings of the IFIP International Conference on Artificial Intelligence in Theory and Practice, Santiago, Chile, 21–24 August 2006; Springer: Boston, MA, USA, 2006; pp. 149–158.
47. Dupret, G.; Koda, M. Bootstrap re-sampling for unbalanced data in supervised learning. *Eur. J. Oper. Res.* **2001**, *134*, 141–156. [CrossRef]
48. Zweig, M.H.; Campbell, G. Receiver-operating characteristic (ROC) plots: A fundamental evaluation tool in clinical medicine. *Clin. Chem.* **1993**, *39*, 561–577. [CrossRef]
49. Baltrušaitis, T. Automatic Facial Expression Analysis. Ph.D. Thesis, University of Cambridge, Cambridge, UK, 2014.
50. Dahliani, E.R.; Rahmatan, H.; Djufri. The correlation between students' interest and learning outcomes in biology. *JPhCS* **2020**, *1460*, 012072. [CrossRef]
51. Widanti, N.; Sumanto, B.; Rosa, P.; Fathur Miftahudin, M. Stress level detection using heart rate, blood pressure, and GSR and stress therapy by utilizing infrared. In Proceedings of the 2015 International Conference on Industrial Instrumentation and Control (ICIC), Pune, India, 28–30 May 2015; pp. 275–279.
52. Lacy, C.R.; Contrada, R.J.; Robbins, M.L.; Tannenbaum, A.K.; Moreyra, A.E.; Chelton, S.; Kostis, J.B. Coronary vasoconstriction induced by mental stress (simulated public speaking). *Am. J. Cardiol.* **1995**, *75*, 503–505. [CrossRef]
53. Kim, H.-Y. Statistical notes for clinical researchers: Assessing normal distribution (2) using skewness and kurtosis. *Restor. Dent. Endod.* **2013**, *38*, 52–54. [CrossRef] [PubMed]
54. Pereira, C.B.; Yu, X.; Czaplik, M.; Rossaint, R.; Blazek, V.; Leonhardt, S. Remote monitoring of breathing dynamics using infrared thermography. *Biomed. Opt. Express* **2015**, *6*, 4378–4394. [CrossRef] [PubMed]
55. Widjaja, D.; Orini, M.; Vlemincx, E.; Van Huffel, S. Cardiorespiratory Dynamic Response to Mental Stress: A Multivariate Time-Frequency Analysis. *Comput. Math. Methods Med.* **2013**, *2013*. [CrossRef] [PubMed]
56. Vlemincx, E.; Taelman, J.; Peuter, S.D.; Diest, I.V.; Bergh, O.V.D. Sigh rate and respiratory variability during mental load and sustained attention. *Psychophysiology* **2011**, *48*, 117–120. [CrossRef]
57. Lohani, M.; Payne, B.R.; Strayer, D.L. A review of psychophysiological measures to assess cognitive states in real-world driving. *Front. Hum. Neurosci.* **2019**, *13*. [CrossRef]

58. Hill, J.D.; Boyle, L.N. Driver stress as influenced by driving maneuvers and roadway conditions. *Transp. Res. Part F Traffic Psychol. Behav.* **2007**, *10*, 177–186. [CrossRef]
59. Matthews, G.; Joyner, L.A.; Newman, R. Age and gender differences in stress responses during simulated driving. *Proc. Hum. Factors Ergon. Soc. Annu. Meet.* **1999**, *43*, 1007–1011. [CrossRef]
60. Daanen, H.A.; Van De Vliert, E.; Huang, X. Driving performance in cold, warm, and thermoneutral environments. *Appl. Ergon.* **2003**, *34*, 597–602. [CrossRef]

applied sciences

MDPI

Article

Driver Facial Expression Analysis Using LFA-CRNN-Based Feature Extraction for Health-Risk Decisions

Chang-Min Kim [1], Ellen J. Hong [2], Kyungyong Chung [3] and Roy C. Park [4,*]

1 Division of Computer Information and Engineering, Sangji University, Wonju 26339, Korea; changingstart@gmail.com
2 Department of Computer and Telecommunications Engineering, Yonsei University, Wonju 26493, Korea; jeonghee.hong@gmail.com
3 Division of Computer Science and Engineering, Kyonggi University, Suwon 16227, Korea; dragonhci@gmail.com
4 Department of Information Communication Software Engineering, Sangji University, Wonju 26339, Korea
* Correspondence: roypark1984@gmail.com

Received: 26 February 2020; Accepted: 22 April 2020; Published: 24 April 2020

check for updates

Abstract: As people communicate with each other, they use gestures and facial expressions as a means to convey and understand emotional state. Non-verbal means of communication are essential to understanding, based on external clues to a person's emotional state. Recently, active studies have been conducted on the lifecare service of analyzing users' facial expressions. Yet, rather than a service necessary for everyday life, the service is currently provided only for health care centers or certain medical institutions. It is necessary to conduct studies to prevent accidents that suddenly occur in everyday life and to cope with emergencies. Thus, we propose facial expression analysis using line-segment feature analysis-convolutional recurrent neural network (LFA-CRNN) feature extraction for health-risk assessments of drivers. The purpose of such an analysis is to manage and monitor patients with chronic diseases who are rapidly increasing in number. To prevent automobile accidents and to respond to emergency situations due to acute diseases, we propose a service that monitors a driver's facial expressions to assess health risks and alert the driver to risk-related matters while driving. To identify health risks, deep learning technology is used to recognize expressions of pain and to determine if a person is in pain while driving. Since the amount of input-image data is large, analyzing facial expressions accurately is difficult for a process with limited resources while providing the service on a real-time basis. Accordingly, a line-segment feature analysis algorithm is proposed to reduce the amount of data, and the LFA-CRNN model was designed for this purpose. Through this model, the severity of a driver's pain is classified into one of nine types. The LFA-CRNN model consists of one convolution layer that is reshaped and delivered into two bidirectional gated recurrent unit layers. Finally, biometric data are classified through softmax. In addition, to evaluate the performance of LFA-CRNN, the performance was compared through the CRNN and AlexNet Models based on the University of Northern British Columbia and McMaster University (UNBC-McMaster) database.

Keywords: facial expression analysis; line segment feature analysis; dimensionality reduction; convolutional recurrent neural network; driver health risk

1. Introduction

In our lives: emotion is an essential means to deliver information among people. Emotional expressions can be classified in one of two ways: verbal (the spoken and written word) and non-verbal (gestures, facial expressions, etc.) [1,2]. People communicate with others every day, and in such a process, facial expressions account for a significantly high proportion of meaning. Expressions can be used to accurately understand another person's emotional state, and since the perception of the emotion in facial expressions is an essential factor of social cognition, it could be seen as playing an essential role in diverse areas of life [3,4]. As described, facial expressions can be used to understand and empathize with other people's emotions. A number of services and prediction models for analyzing such expressions and understanding users' emotional states have been, and are being, studied [5,6]. Recently studied are lifecare services using gesture recognition or expression analysis to detect the risks to which the elderly and patients with chronic diseases are exposed. Currently, the whole world is entering an aging society, and accordingly, the number of patients with chronic diseases (hypertension, cardiovascular diseases, and coronary artery disease, etc.) is increasing. In addition, even in the low age group, the prevalence of chronic diseases increases due to changes in dietary habits (food with high calories and high sugar content, etc.), lack of exercise, and smoking, etc. [7,8]. From one generation to the next, humankind will require services to continuously monitor and manage chronic diseases. Unless medical service technology makes innovative progress, the demand for such services will continue to increase. There are many cases where the elderly or patients with diseases experience emergencies, major accidents, or death from disease-related reasons such as acute shock. In particular, since car accidents frequently occur due to acute diseases while someone is driving, it is necessary to take urgent action to prevent them [9–11]. When such accidents occur in the absence of a fellow passenger, the driver is unable to take prompt action. Moreover, as autonomous vehicles become more popular, they can operate on cruise control regardless of the status of the driver. In such cases, even after a driver's abnormal health status is detected, he or she might not be able to do something within the so-called "golden time" to address the problem. Although the mortality of patients with chronic diseases has decreased due to medical progress, it is still necessary to continuously manage and prepare for emergency situations [12,13]. To that end, services are being studied that alert friends, hospitals, police stations, etc., after detecting a driver at risk. However, since such studies involved prediction models based on external factor analysis, a driver's potential risk factors have not been applied, and it is difficult to predict accidents that occur due to internal factors. For prediction services in which internal factors apply, accurate prediction requires a particular device to be installed in the vehicle, and a given code of conduct is to be followed. As for the possibility of checking the driver's internal risk factors intuitively, it is possible to judge a dangerous situation through the driver's facial expressions. Thus, this study was conducted for the prediction of risk through recognition of the driver's facial expressions. Concerning current recognition of facial expressions, various studies are in progress [14]. As for the traditional face recognition techniques, classification models using the extraction of handcrafted features (local binary pattern (LBP), histogram of gradients (HOG), gabor, scale-invariant feature transform (SIFT)) for the sensible extraction of the characteristics of face images have usually been used [15–17]. These methods have a problem, however, in which the performance deteriorates when there are various changes in the face in an actual environment. It is difficult to choose the appropriate feature parameters according to the field of application. Transformation can be made in various shapes, but it is necessary to determine the optimal feature parameters via experiential elements and various experiments, which is a problem. Recently, the deep-learning technique has widely been used. As the face recognition technique based on deep learning itself learns high-level characteristics, using a large amount of data built up in various environments, it shows a high recognition performance even in a wild environment. Accordingly, DeepFace (based on AlexNet) uses the locally connected convolution layer to effectively extract local characteristics from the face region [18]. DeepID proposed a lightweight convolutional neural network (CNN)-based face recognizer, using an input resolution with smaller pixels than DeepFace [19]. VGGFace, which appeared later, learned a deep network structure consisting of 15 convolution layers using a data set for high-capacity face recognition made by itself through an Internet search [20]. In addition, various studies were conducted to improve the performance of face recognition models such as DeepID2, DeepID3, and GoogLeNet [21–23]. As yet, if real-time video data are processed by a face recognition technique based on deep learning, it is necessary to learn many classes. Thus, they mostly show the structure in which a fully connected layer becomes larger, which accordingly decreases the batch size and acts as a factor disturbing convergence in the learning by the neural network. Accordingly, in this paper, to resolve such problems, facial expression analysis of drivers by using line-segment feature analysis-convolutional recurrent neural network (LFA-CRNN) feature extraction for health-risk assessment is proposed. A service using facial expression information to analyze drivers' health risks and alert them to risk-related matters is proposed. Drivers' real-time streaming images, along with deep learning-based pain expression recognition, were utilized to determine whether or not drivers are suffering from pain. When analyzing real-time streaming images, it may be difficult to extract accurate facial expression features if the image is shaking, and it may be difficult or impossible to run the analysis process on a real-time basis due to limited resources. Accordingly, a line-segment feature analysis (LFA) algorithm reduces learning and assessment time by reducing data dimensionality (the number of pixels). Also proposed is increasing the processing speed to handle large-capacity original data and high resolutions. Drivers' facial expressions are recognized through the CRNN model, which is designed to reduce input data dimensionality and to learn the LFA data. The driver's condition is understood based on the University of Northern British Columbia and McMaster University (UNBC-McMaster) database to understand the driver's abnormal condition. A service is proposed for coping with risks, spreading the dangerous conditions concerning health risks that may occur while driving through the notice by understanding the driver's conditions as suffering and non-suffering conditions.

This study is organized as follows. Section 2 presents the trends in face analysis research and also describes the current risk-prediction systems and services using deep learning. Section 3 describes how the dimensionality-reducing LFA technique proposed in this paper is applied to the data generation process, and also presents the CRNN model designed for LFA data learning. Section 4 describes how the UNBC-McMaster database was used to conduct a performance test.

2. Related Research

2.1. Face Analysis Research Trends

In early facial expression analysis, various studies were conducted based on a local binary pattern (LBP). LBP is widely used in the field of image recognition thanks to its ability to recognize things, its strength against changes in lighting, and its ease of calculation. As LBP became widely used in face recognition, center-symmetric LBP (CS-LBP) [24] was used in a modified form that can show components in the diagonal direction, reducing the dimension of feature vectors. Also, some studies enhanced the accuracy of facial expression detection by using multi-scale LBP that multiplied the size of the radius and the angle [25,26]. However, the LBP technique is used with techniques for extracting feature vectors in order to increase accuracy. In this case, based on the field of the application, there is difficulty in choosing the appropriate feature vectors. Transformation in various forms is possible, but the optimal feature vector should be decided by experiential elements and from various experiments. If the LFA proposed in this study is used, the minimum necessary data are used when the face is analyzed, so data compression takes place autonomously. Also, since it can be performed through techniques for detecting the face and its outline, it can easily be used in various fields. Studies of face analysis based on point-based features utilizing landmarks are also in progress. Landmark-based face extraction has a very fast process of measuring and restoring the landmark, so it can immediately display changes in the face shape and facial expressions filmed in real time. The weight of the measured landmark can be lightened for uses and purposes, such as character and avatar. Jabon et al. (2010) [27] proposed a prediction model that could prevent traffic accidents by recognizing drivers' facial expressions and gestures. This prediction model generates 22 x and y coordinates on the face (eyes, nose, mouth, etc.) in order to extract facial characteristics and head movements, and it automatically detects movement. It synchronizes the extracted data with simulator data, uses them as input to the classifier, and calculates a prediction for accidents. Also, Agbolade et al. (2019) [28] and Park (2017) [29] conducted studies to detect the face region based on multiple points, utilizing landmarks to increase the accuracy of face extraction. However, to prevent prediction of the landmark value from falling to the local minimum, it is necessary to pass through a process of correcting the result through plural networks based on the initial prediction value in cascade form. The difficulty in detection differs depending on the set value of the feature point of the face. The more subdivided the overall detected outline, the more difficult it gets. Also, if part of the face is covered, it becomes very hard to measure landmarks. If the LFA proposed in this study is used, it is somewhat possible to escape the impact of light, since only information about the segments is used. Also, there is no increase in the difficulty of detection.

Since the deep learning method shows high performance, studies based on CNNs and deep neural networks (DNNs) are actively conducted. Wang et al. (2019) [30] proposed a method for recognizing facial expressions by combining extracted characteristics with the C4.5 classifier. Since some problems still existed (e.g., overfitting of a single classifier, and a vulnerable generalization ability), ensemble learning was applied to the decision-making tree algorithm to increase classification accuracy. Jeong et al. (2018) [31] detected face landmarks through a facial expression recognition (FER) technique proposed for face analysis, and extracted geometric feature vectors considering the spatial position between landmarks. By implementing the feature vectors on a proposed hierarchical weighted random forest classifier in order to classify facial expressions, the accuracy of facial recognition increased. Ra et al. (2018) [32] proposed a deep learning structure in a block method to enhance the face recognition rate. Unlike the existing method, feature filter coefficients and the weighted values of the neural network (on the softmax layer and the convolution layer) are learned using a backpropagation algorithm. Performing recognition with the deep learning model that learned the selected block region, the result of face recognition is drawn from an efficient block with a high feature value. However, since the face recognition technique based on CNNs and DNNs should generally learn a large amount of classes, there is a structure in which the fully connected layer grows bigger. Accordingly, the structure acts as a factor reducing the batch size and disturbing convergence in the learning by a neural network. If the LFA proposed in this study is used, the input dimension is small. Thus, the disturbance in the convergence from learning (due to the decrease in the batch size that may be generated in the CNN and DNN) can be minimized.

2.2. *Facial Expression Analysis and Emotion-Based Services*

FaceReader automatically analyzes 500 features on a face from images, videos, and streaming videos that include facial expressions, and it analyzes seven basic emotions: neutrality, happiness, sadness, anger, amazement, fear, and disgust. It also analyzes the degree of the emotions, such as the arousal (active vs. passive) and the valence (positive vs. negative) online and offline. Research on emotions through analyzing facial expressions has been conducted in various research fields, including consumer behavior, educational methodology, psychology, consulting and counseling, and medicine for more than 10 years. It is widely used in more than 700 colleges, research institutes, and companies around the world [33]. The facial expression-based and bio-signal-based lifecare service provided by Neighbor System Co. Ltd. in Korea is an accident-prevention system dedicated to protecting the elderly who live alone and who have no close friends or family members. The services provided by this system include user location information, health information confirmation, and integrated situation monitoring [34]. Figure 1 shows the facial expression-based and bio-signal-based lifecare service, which consists of four main functions for safety, health, the home, and emergencies.

The safety function provides help/rescue services through tracing/managing the users' location information, tracing their travel routes, and detecting any deviations from them. The health function measures/records body temperature, heart rate, and physical activity level, and monitors health status. In addition, it determines whether or not an unexpected situation is actually an emergency by using facial expression analysis, and provides services applicable to the situation. The home function provides a service dedicated to detecting long-term non-movement and to preventing intrusions by using closed-circuit television (CCTV) installed within the users' residential space. Lastly, the emergency function constructs a system with connections to various organizations that can respond to any situation promptly, as well as deliver users' health history records to the involved organizations.

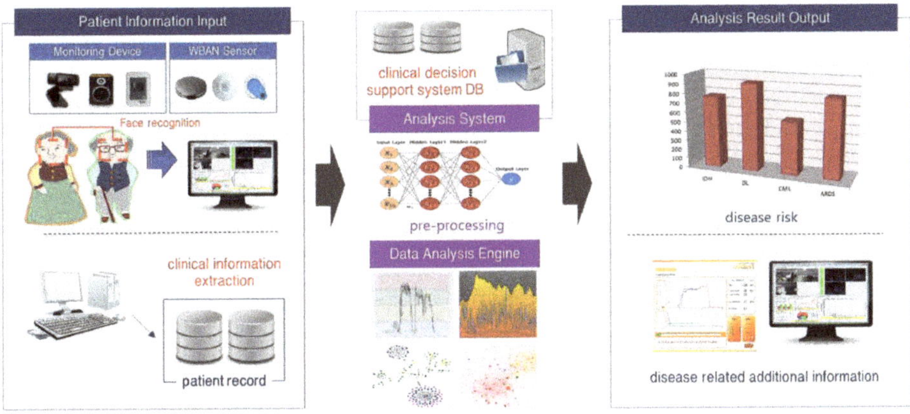

Figure 1. Facial expression and bio-signal-based lifecare service.

3. Driver Health-Risk Analysis Using Facial Expression Recognition-Based LFA-CRNN

It is necessary to compensate for senior drivers' weakened physical, perceptual, and decision-making abilities. It is also necessary to prevent secondary accidents, manage their health status, and take prompt action by predicting any potential traffic-accident risk, health risk, and risky behavior that might show up while driving. In cases where a senior driver's health status worsens due to a chronic disease, it becomes possible to recognize accident risks through facial expression changes. Accordingly, we propose resolving such issues with facial expression analysis using LFA-CRNN-based feature extraction for health-risk assessment of drivers. The LFA algorithm was performed to extract the characteristics of the driver's facial image in real time in the transportation support platform. An improved CRNN model is proposed, which can recognize the driver's face through the data calculated in this algorithm. Figure 2 shows the LFA-CRNN-based driving facial expression analysis for assessing driver health risks.

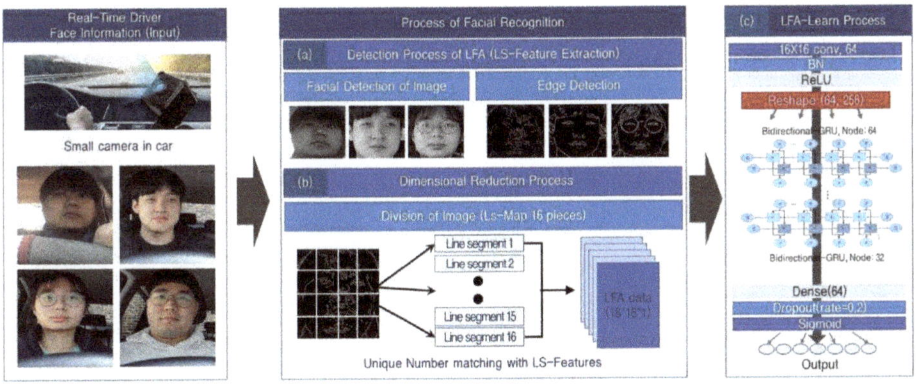

Figure 2. Line-segment feature analysis-convolutional recurrent neural network (LFA-CRNN)-based facial expression analysis for driver health risk assessment.

The procedures for recognizing and processing a driver's facial expressions can be divided into detection, dimensionality reduction, and learning. The detection process is a step of extracting the core areas (the eyes, nose, and mouth) to analyze the driver's suffering condition. In the step, there is a preconditioning process to solve the problem that the core areas are not accurately recognized. To extract features from the main areas of frame-type facial images segmented from real-time streaming images, multiple AdaBoost-based input images are divided into blocks. In the dimensionality reduction process, the LFA algorithm reduces the learning and reasoning time by reducing data dimensionality (the number of pixels) in order to increase processing speeds to handle large-capacity original data. High-resolution data and the dimensionality of the input data are reduced. Lastly, in the learning process, drivers' facial expressions are recognized through the CRNN model designed to learn the LFA data. In addition, to confirm a driver's abnormal status based on the UNBC-McMaster shoulder pain expression database, the service proposed determines if the driver is in pain, identifies the driver's health-related risks, and alerts the driver to such risks through alarms.

3.1. Real-Time Stream Image Data Pre-Processing for Facial Expression Recognition-Based Health Risk Extraction

Because pre-existing deep-learning models utilize the overall facial image for facial recognition, areas such as the eyes, nose, and lips serving as the main factors for analyzing drivers' emotions and pain status are not accurately recognized. Accordingly, through a detection process module, pre-processing is conducted for dimensionality reduction and learning. To analyze the original data transferred through real-time streaming, input images are segmented at 85 fps, and to increase the recognition rate, the particular facial image sections required for facial expression recognition are extracted using the multi-block method [35]. In particular, in cases where a multi-block is big or small during the blocking process, pre-existing models are unable to accurately extract features from the main areas, and this causes significant errors relating to recognition and learning. To resolve such issues, multiple AdaBoost is utilized to set optimized blocking, and then sampling is conducted. Figure 3 shows the process of detecting particular facial areas. A Haar-based cascade classifier is used to detect the face; Haar-like features are selected to accurately extract the user's facial features, and the AdaBoost algorithm is used for training. At this point, since features can be seen as a face/background-dividing characteristic and as a classifier, each feature is defined as a base classifier or a weak classifier candidate. During iterations, the training samples select one feature demonstrating the best classification performance, and the selected feature is used as the weak classifier in the iteration. The final weak classifiers are used in the weighted linear combination process to acquire the final strong classifiers.

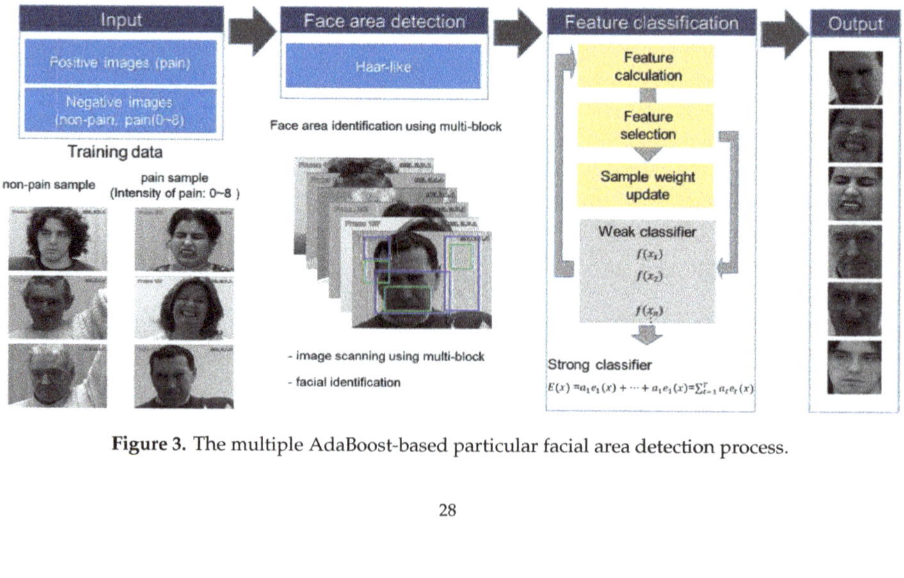

Figure 3. The multiple AdaBoost-based particular facial area detection process.

In the formula in Figure 3, $E(x)$ is the strong classifier finally found; e is the weak classifier drawn in the learning process, and a is the weighted value for the weak classifier. T is the number of repetitions. In this process, it is very hard to normalize the face if it is extracted without information such as the rotation and position of the face. Extracting the geometrical information of the face, it is necessary to normalize the face consistently. Faces can be classified according to their rotational positions, and if random images do not provide such information in advance, such rotational information must be detected during image retrieval. The detectors learned through multiple Adaboost are serialized, using the simple pattern of the face searcher. Using the serialized detectors, information can be found, such as the position, size, and rotation of the face. As for the simple pattern used in multiple Adaboost learning, the pattern in a basic form was used. 160 was chosen as the number of simple detectors to be found by Adaboost learning, and the processing speed of the learned detectors improved through serialization. The face region calculated through the above process detects the outline of the face through the Canny technique. This is the optimal technique option based on the experimental result. In the early stages, various outline detection techniques were used, but only the Canny method showed a high result.

3.2. Line-Segment Feature Analysis (LFA) Algorithm for Real-Time Stream Image Analysis Load Reduction

3.2.1. Pain Feature Extraction through LFA

Even after executing facial feature extraction through the procedures specified in Section 3.1, various constraint conditions may arise when extracting a driver's facial features from real-time driving images. In analyzing a real-time streaming image, it may be hard to extract accurate facial characteristics due to the motion of the image. Accordingly, since it is necessary to reduce the dimensionality of facial feature images extracted from real-time streaming images, the LFA algorithm is proposed. The proposed LFA algorithm is a dimensionality-reduction process that reduces learning and reasoning time by reducing data dimensionality (the number of pixels) to increase the processing speed in order to handle the original large-capacity, high-resolution data. To extract information from images, the line information on a 3×3 Laplacian mask's parameter-modified filter is extracted, a one-dimensional (1D) vector is created, and the created vector is utilized as the learning model's input data. Based on such a process, this algorithm creates new data through the line-segment features. LFA uses the driver's facial contour lines calculated through the detection process to examine and classify line-segment types. To examine the line-segment types, a filter, f, is used, and the elements {1, 2, 4, and 8} are acquired. Figure 4 shows the process where a driver's facial-contour line data are segmented, and the line-segment types are examined through the use of f.

Algorithm 1 Image Division Algorithm

```
Input: [x_1, x_2, ... , x_n]
def Division and Max-pooling of image
        Y = List()
        for x_i in [x_1, x_2, ... , x_n] do
              sub = List()
              for w from 0 to D_w do // D_w, D_h denote the size of the image to be divided.
                 for h from 0 to D_h do
                        // fw, fh denote the size of the filter.
                        sub.append(x_i[w*fw: (w + 1)*fw, h*fh: (h + 1)*fh])
              Y.append(sub)
           Y = Max-pooling(Y, stride = (2,2), padding = 'same')
Output: Y[Y_1, Y_2, ... , Y_n]
```

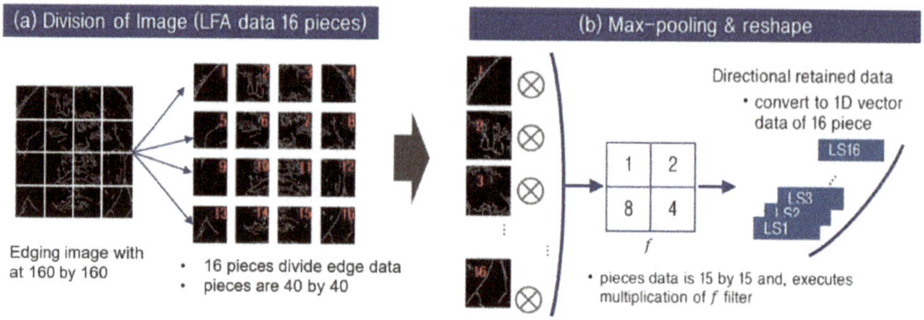

Figure 4. Driver facial contour-line data segmentation and line-segment type examination using f: (**a**) Division of image (LFA data 16 pieces); (**b**) Max-pooling and reshape.

Figure 4 shows the first LFA process. The contour line image calculated through pre-processing (detection) had a size of 160 × 160, and this image was segmented into 16 parts, as shown in Figure 4a. This process is calculated as shown in Algorithm 1. The segmented parts have a size of 40 × 40, and the segments are arranged in a way that does not modify the structure of the original image. These segments are max-pooled via the calculation shown in Figure 4b, and the arrangement of the segments is adjusted. This process is defined in Equation (1):

$$D_w, D_h = 4, 4, P_w = \frac{W}{D_w}, P_h = \frac{H}{D_h} P[\mathrm{n,m}] = x[n * P_w : (n+1) * P_w, m * P_h : (m+1) * P_h], \\ (0 \le \mathrm{n} \le 4, 0 \le \mathrm{m} \le 4) MP[\mathrm{n,m}] = max - pooling(P[n,m]), \quad (1)$$

Equation (1) is a calculation where the contour line image obtained during pre-processing is divided into 16 equal segments, and the divided segments are max-pooled. D_w and D_h denote the number of segments in the width and height, respectively, and P_w and P_h denote the size of the segmented data from dividing the contour line image by D_w and by D_h, respectively. P indicates the space for memorizing the segmented data, and the segmentation position is maintained through P[n, m], in which n and m refer to a two-dimensional array index, having a value ranging between 0 and 4. MP memorizes the segmented data's max-pooling results. In every process, the sequence of the segmented images must not be lost. The sequence of the re-arranged segments must not be lost as well. Figure 4b shows the calculation where a convolution between the segment images and the filter is calculated: the parameters of the segmented images are converted, the sum of the parameters is calculated, and one-dimensional vector data are generated. The number of segmentations and the size of images in this process were the values selected through experiential selection, and after experimenting on various conditions, the optimal variables were calculated.

3.2.2. Line-Segment Aggregation-Based Reduced Data Generation for Pain Feature-Extracted Data Processing Load Reduction

The information from the line segment (LS) extracted (based on real-time streaming images) is matched with a unique number. The unique numbers are 1, 2, 4, and 8; they have a value that does not overlap another value, and the aggregate value deduces mutually different values. The LFA algorithm uses a 2 × 2 filter having a unique number for matching normal line-segment data. The LS has a value of 0 or 1, and where a filter consisting of a unique number is matched with the LS, only the areas having 1 as the unique number are displayed. A serial number is given to express information on segments, which is visual data in a series of information on numbers. That is converted to a series of information on numbers for easy counting of various segments (curve, horizontal line, and vertical line, etc.). Namely, visual data are converted into a series of patterns (numbers). Figure 5 shows the process where a segmented image is converted into 1D vector data.

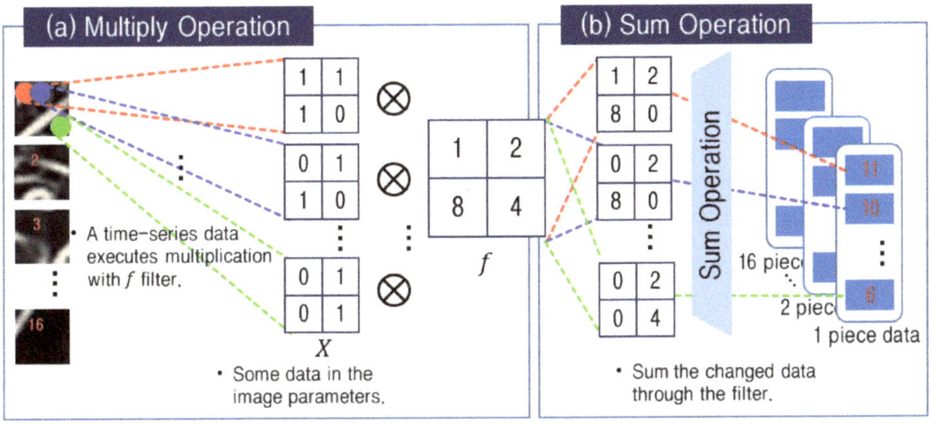

Figure 5. Conversion from a segment image to 1D vector data: (**a**) Multiply operation; (**b**) Sum operation.

A segment of an image utilizing contour line data has a parameter of 0 or 1, as shown in Figure 5a. The involved segment is a line segment when this parameter is 1, and is background when the parameter is 0. Such segment data are calculated with the filter, f, in sequence. The segment data have a size of 20 × 20, and filter f is 2 × 2. The 2 × 2 window is used to calculate a convolution between the segment data and filter f. At this point, the window moves one pixel at a time (stride = 1) to scan the entire area of the segmented image. Each scanned area is calculated with filter f; the parameter is changed, and the image's 1 parameter is replaced with the f parameter. The process in Figure 5b is calculated as shown in Algorithm 2.

Algorithm 2 1D Vector Conversion Algorithm

```
Input: [x1 = [p1, p2, ... , p16], x2 = [p1, p2, ... , p16], ... , xn = [p1, p2, ... , p16]]
def Convert image to a 1D vector
    Label = [1, 2, 8, 4]
    Y = List()
    for xi in [x1, x2, ... , xn] do
    // Sub1 is a list to save the result of a piece of the image.
      sub1 = List ()
      for pi in xi do
      // Sub2 is a list to save the result of the image of the matched piece
      // with label data.
        sub2 = List ()
        for w from 0 to W-fw+1 do
          for h from 0 to H-fh+1 do
            p = pi[w:w + fw, h:h + fh]
            p = p.reshape(−1) * Label
            sub2.append(sum(p))
        sub1.append(sub2)
      Y.append(sub1)
Output: Y[Y1, Y2, ... , Yn]
```

Equation (2) shows the calculation between the segment image and filter f, in which f_w and f_h represent the size. At this point, f has a fixed parameter and a fixed size; x_i is a partial area of the segment image, and segments it into pieces the same size as f. Once a convolution between such segmented data and f is calculated, the calculated results are added up and recorded in P_i.

$$f = [[1,2],[8,4]] f_w, f_h = 2, 2 P_i = \sum_{w=0}^{W} \sum_{h=0}^{H} x_i[w : w + f_w, h : h + f_h] \otimes f, \quad (2)$$

For example, when the segment image has parameters set to [[1, 0], [0, 1]], as a convolution between the segment image and f is calculated, the segment image's parameters are changed to [[1, 0], [0, 4]]. These changed parameters obtain different values according to the position of each parameter due to f. Adding them up shows a different result, according to the data expression of the scan area. Table 1 shows the type and sum of lines according to the scanned areas.

Table 1. Type and sum of lines according to the scanned areas.

Scanned Area	Summing Data	Line Type	Scanned Area	Summing Data	Line Type
0000	0	Non-Active	1000	8	Point
0001	1	Point	1001	9	Vertical
0010	2	Point	1010	10	Diagonal
0011	3	Horizontal	1011	11	Curve
0100	4	Point	1100	12	Horizontal
0101	5	Diagonal	1101	13	Curve
0110	6	Vertical	1110	14	Curve
0111	7	Curve	1111	15	Active (Side)

When scanned areas are expressed as 0, they are considered the background (as shown in Table 1), and the summed value is also expressed as 0. On the other hand, all the areas expressed as 1 are considered active, and the side acquires a value expressed as 15. Other areas, according to the position and number of 1s, are expressed as point, vertical, horizontal, or diagonal, and are given a unique number. Despite being identical line types, all data are assigned a different number according to the expressed position, and the summed value is the unique value. For example, vertical is one of the line types detected in areas expressed as 0110 or 1001. However, each summed value is either 6 or 9 and has a different unique value. This means that the same line types are considered different lines based on their line-expressed positions. In addition, each line type's total cannot exceed 15. The data calculated through such a process will tie and save the line types (total) calculated per segment as a 1D vector, and will create a total of 16 1D vectors. Each vector has a size of $(20 - 2 + 1) \times (20 - 2 + 1) = 841$, and each vector's parameter has a value ranging from 0 to 15.

3.2.3. Unique Number-Based Data Compression and Feature Map Generation for Image Dimensionality Reduction

The 16 one-dimensional vector data calculated through the process shown in Figure 5 consist of unique values according to the line type determined through the information calculated by segmenting the facial image into 16 parts and matching each part with a particular filter. Such vector data consist of parameters ranging from 0 to 15. Each parameter has a unique feature (line-segment information). This section describes how cumulative aggregate data are generated based on the parameter value owned by each segment. The term "cumulative aggregate data" refers to data generated through a process where a parameter value is utilized as an index to generate a 1D array having a size of 16. The involved array's factor increases by 1 every time each index is called. Figure 6 shows the process where cumulative aggregate data are generated.

Algorithm 3 Cumulative Aggregation Algorithm

Input: $[x_1 = [p_1 = [v_1, v_2, \ldots v_m], p_2, \ldots, p_{16}], x_2, \ldots, x_n]$
def Cumulative aggregation used to make LFA data
  Y = List()
  for x_i in $[x_1, x_2, \ldots, x_n]$ **do**
    sub1 = List()
    for p **in** x_i **do**
      sub2 = array(16){0, … }
      for i **from** p **do**
        sub2[i]++
      sub1.append(sub2)
    Y.append(sub1)
Output: Y

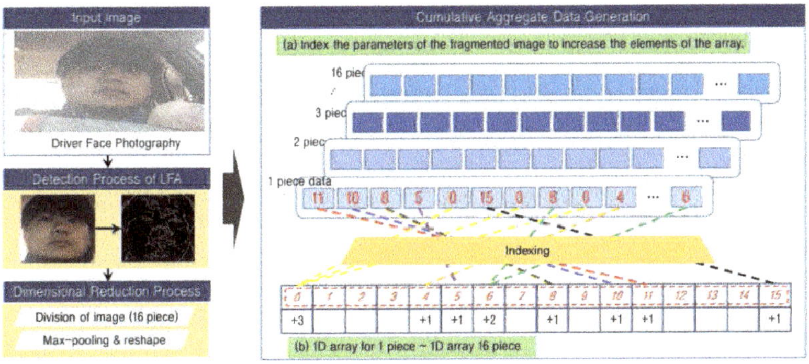

Figure 6. Process of cumulative aggregate data generation: (**a**) Index the parameters of the fragmented image to increase the elements of the array; (**b**) 1D array for 1 piece~1D array 16 piece.

As shown on the right side of Figure 6a, the parameters of the data segmented through the previous process are utilized as an array of the index, and a 1D array having a size of 16 is generated according to each segment. This array is shown in Figure 6b, and the factor value of the index position corresponding to each parameter of the one-dimensional array having a maximum size of 16 increases by 1. The process in Figure 6b is calculated as shown in Algorithm 3. Since this process is applied to each segment, an array having a size of 16 and corresponding to each segment is generated for each segment, and a total of 16 arrays are generated. These are known as LFA data and are shown in Figure 7a.

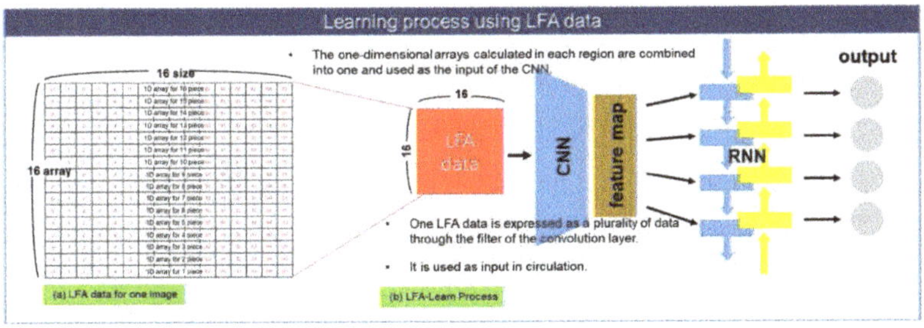

Figure 7. Learning process using LFA data: (**a**) LFA data for one image; (**b**) LFA-Learn process.

The LFA process in Figure 7a restructures each array generated through each segment image in the appropriate order (in the order prioritized based on the segmentation position). Through this, for one image, the LFA data calculated through the LFA process are expressed as two-dimensional sequences with a size of 16 × 16. This is used as input for the CRNN.

3.3. LFA-CRNN Model for Driver Pain Status Analysis

Once a feature map is generated, and the image is deduced through facial and contour line detection, the pre-processing of the given input images restructures them into two-dimensional arrays having a size of 16 × 16 through the LFA process. Specifically, the dimensionality is reduced through the LFA technique. Since LFA always has the same output size and consists of aggregate information on the line segment contained in the image, the reduced data themselves can be considered unique features. In addition, a learning model dedicated to LFA data is designed instead of a general CRNN learning model architecture for drivers' pain status, and the learning process is performed as well. Figure 8 shows the structure of the proposed LFA-CRNN model.

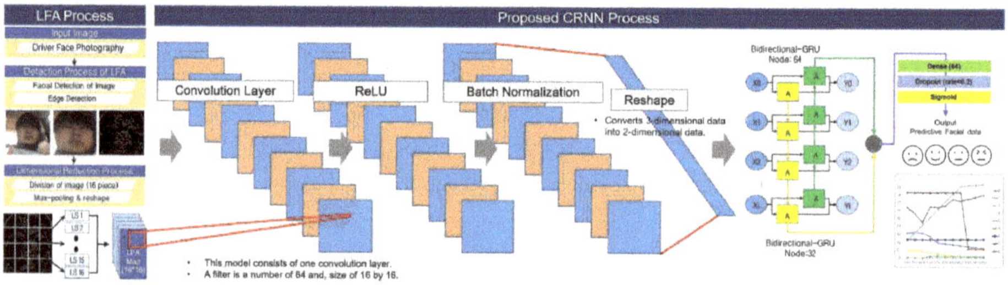

Figure 8. The proposed LFA-CRNN architecture.

The LFA-CRNN architecture is a CRNN learning model. It consists of one convolution layer, and expresses a feature map as sequence data through the reshaped layer. The features that changed into sequence data are transferred to the dense layer through two bidirectional gated recurrent units (BI-GRUs), and the sigmoid layer serves as the final layer before the results are output. Through the convolution layer's batch normalization (BN), the risk of depending on and overfitting the learning-speed improvement and initial weighted-value selection is reduced [36–38]. Since this learning model uses dimensionality-reduced LFA data, the compressed data themselves can be considered one feature. Accordingly, to express one major feature as a number of features in a convolution, the input-related expressions are diversely divided through a total of 64 filters having a size of 16 × 16. The value deduced through such a process passes through the BN and generates a series of feature maps through the rectified linear unit (ReLU) layer. Such feature maps are restructured through the reshape layer into 64 sequence data having a size of 256 and are used as the RNN model's input. The RNN model consists of two BI-GRUs, one with 64 nodes and one with 32 nodes. The data deduced through this process are delivered to the sigmoid layer through the dense layer. At this point, the dropout layer is arranged between the dense layer and the sigmoid layer to prevent calculation volume reduction and overfitting [39–41]. Lastly, through the Sigmoid class, nine types of pain are classified. In this model, the pooling layer generally used in the pre-existing CNN and CRNN models is not used. Since the input LFA data themselves have a considerably small size of 16 × 16, and consist of the cumulative number of line segments owned by the images when the involved data are compressed, the main features may be damaged or removed. In addition, in this model, BN and the dropout layer are arranged instead of the pooling layer, and the convolution's stride and padding are set to *1* and *same*, respectively. We used the convolution layer to get a variety of information about the expression of individual, highly-concentrated LFA data by designing the model like Figure 9. Thus, the filter of the convolution layer was set to 16 × 16 with stride = 1 and padding = "same." Through this, one LFA data size is maintained, and because of the weighted value of the filter, it can express a lot of information. The data are used as input in each cycle of the RNN, and through the previous characteristics, strong characteristics are gradually detected from within.

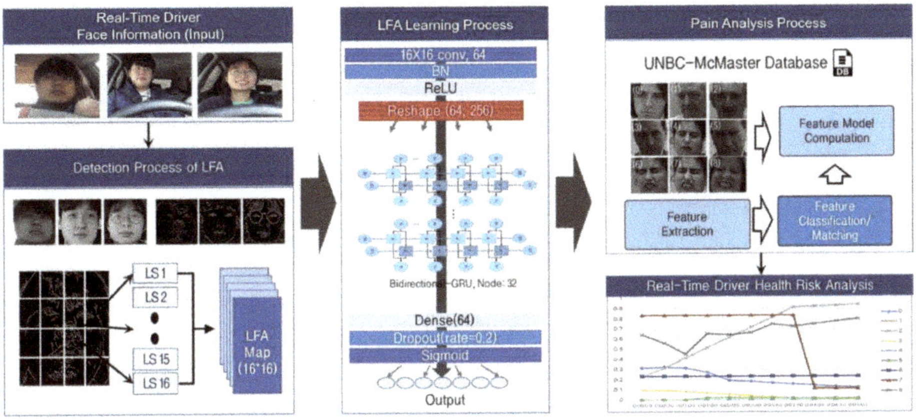

Figure 9. Driver pain status analysis process and its performance evaluation.

4. Simulation and Performance Evaluation

A simulation was conducted in the following environment: a Microsoft Windows 10 pro 64-bit O/S on an Intel Core(TM) i7-6700 CPU (3.40 GHz) with 16GB RAM, and an emTek XENON NVIDIA GeForce GTX 1060 graphics card with 6GB of memory. To implement this algorithm, we utilized OpenCV 4.2, Keras 2.2.4, and the Numerical Python (NumPy) library (version 1.17.4) based on Python 3.6. OpenCV was used to perform the Canny technique during pre-processing by the LFA, and the calculation of the queue generated in the LFA process was performed using the NumPy library. The neural network model was implemented through Keras. Figure 9 shows the process by which the driver's pain status is analyzed and under which the system's performance was evaluated.

To evaluate the performance of the LFA-CRNN model-based face recognition (suffering and non-suffering expressions), the UNBC-McMaster database was used. In addition, a comparison was made with AlexNet and CRNN Models. The experiment of this paper chose the basic structure of the proposed model, the CRNN model and the AlexNet model generally well known for image classification to compare the performance. The UNBC-McMaster database classifies pain into nine stages (0~8) using the Prkachin and Solomon Pain Intensity (PSPI) scale, with data consisting of 129 participants (63 males and 66 females). The accuracy and loss measurement test were based on such data, calculated through pre-processing (face detection and contour line extraction). The LFA conversion process was used as the LFA-CRNN's input, and the CRNN [42] and AlexNet [43] for performance comparison used the data calculated through the face detection process. The test was conducted by taking 20% of the data from the UNBC-McMaster database [44] as the test data, and utilizing 10% of the remaining 80% as verification data. In the process of classifying data, to prevent data from leaning too much towards a particular class, the classification was undertaken by designating a specific percentage for each class. Specifically, 42,512 data units consisted of 29,758 learning data units, 3401 verification data units, and 8503 test data units.

Figure 10 shows the results of the accuracy and loss, using the UNBC-McMaster Database. As shown in Figure 10, the LFA-CRNN showed the highest accuracy, with AlexNet second and the CRNN third. AlexNet showed a large gap between the training data and verification data. The CRNN showed a continuous increase in the training data accuracy but showed a temporary decrease in the verification data accuracy due to overfitting. Although the LFA-CRNN proposed in this paper showed a bit of a gap between the learning and validation data, such a gap is not considered significant. Since no temporary decrease was shown in the validation data, it was confirmed that no learning overfitting occurred; loss data showed the same patterns. AlexNet showed the highest gap between learning and validation data, in terms of loss. The CRNN showed a continuous decrease of loss in both learning and validation data, but showed a temporary increase in validation data. Therefore, the LFA-CRNN can be considered more reliable than both AlexNet and traditional CRNN.

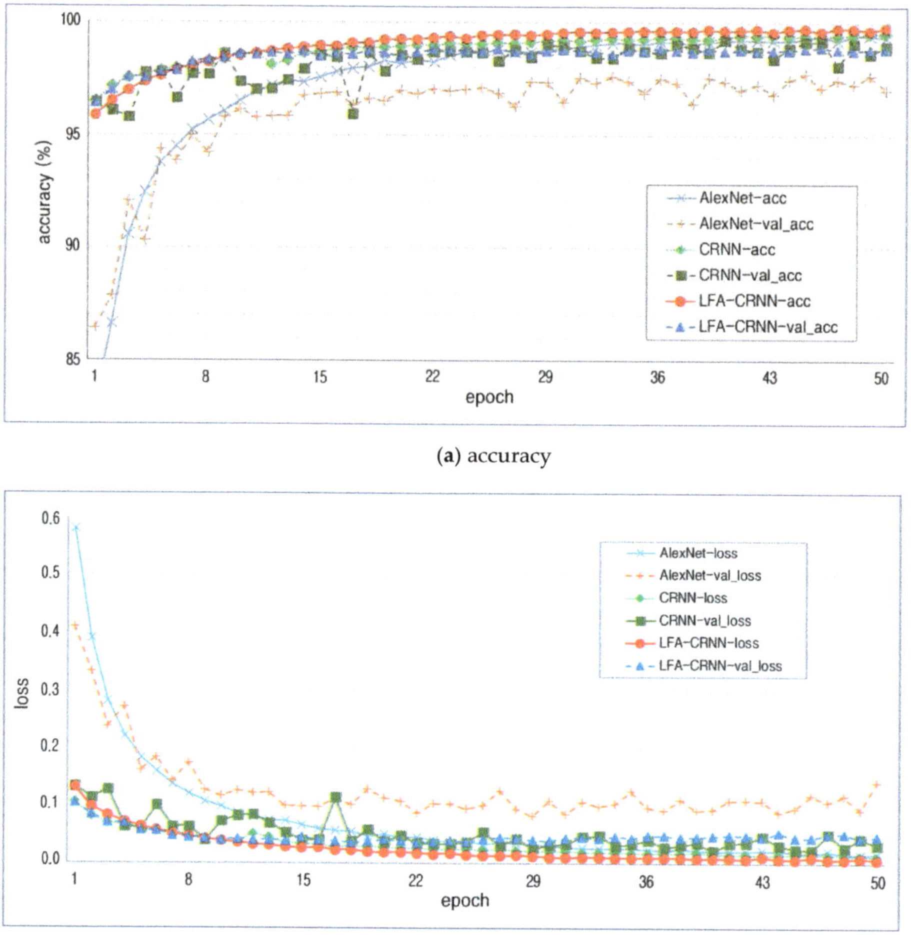

(**a**) accuracy

(**b**) loss

Figure 10. Accuracy and loss measurement results using the University of Northern British Columbia (UNBC)-McMaster shoulder pain expression database.

Figure 11 shows the accuracy and loss achieved with the test data. As shown in the figure, the LFA-CRNN had the highest accuracy at approximately 98.92% and the lowest loss at approximately 0.036. The CRNN showed temporary overfitting during learning, and this was determined to be the reason why its accuracy was lower than the LFA-CRNN. Likewise, it was determined that AlexNet showed a performance decrease in its accuracy due to the verification data's wide gap. The test results shown in Figures 10 and 11 can be summarized as follows. As far as UNBC-McMaster-based learning is concerned, the LFA-CRNN model showed no rapid change in accuracy and loss, and it was confirmed that a stable graph was maintained as the epochs progressed (i.e., no overfitting or large gap). In addition, compared to the basic models, the proposed method showed the highest performance with an accuracy of approximately 98.92%.

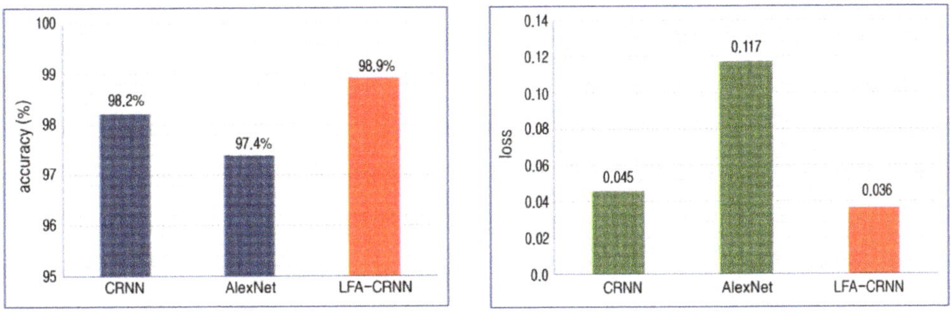

(**a**) Accuracy (**b**) Loss

Figure 11. Accuracy and loss with the test data.

To measure the accuracy and reliability of the proposed algorithm, precision, recall, and the receiver operating characteristic (ROC) curve [45] were measured. Figure 12 shows the results achieved.

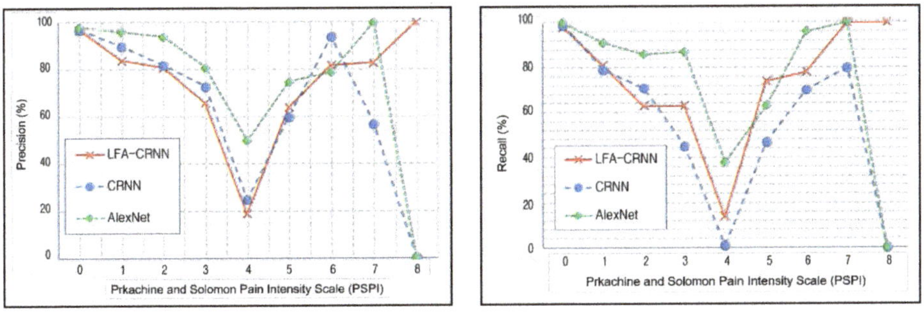

(**a**) Precision (**b**) Recall

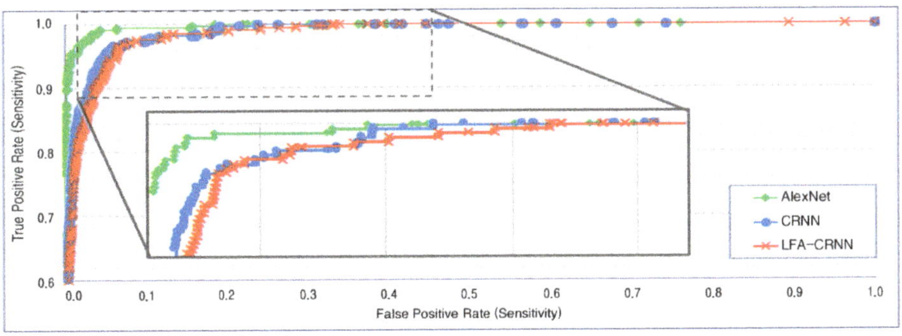

(**c**) ROC curves

Figure 12. Results of precision and recall, plus the receiver operating characteristic (ROC) curve evaluation for each algorithm.

In Figure 12, the precision results show the percentage of the number of samples actually determined to be true out of the samples predicted to be true for each pain severity class. The

LFA-CRNN showed the following results: 0 = 98%, 1 = 81%, 2 = 63%, 3 = 63%, 4 = 19%, 5 = 74%, 6 = 78%, 7 = 100%, and 8 = 100%. Such results are quite poor, compared to the results achieved by AlexNet and the CRNN. It was determined that such results are attributable to the dimensionality reduction LFA technique. Since the dimensionality reduction technique itself either compresses the original image to generate new data or reduces the data size by using particular features consisting of strong features, it removes specific features and only uses strong features. However, only the LFA-CRNN was able to detect data having a PSPI of 8. In addition, as a result of confirming the average precision, both LFA-CRNN and AlexNet showed an average precision of 75%, while the CRNN showed an average precision of 56%. In addition, the recall measurements were similar to the precision results. The LFA-CRNN showed an average recall of 75%, AlexNet showed an average recall of 73%, and the CRNN showed an average recall of 56%. Based on this test, it was confirmed that it was difficult for all the models to detect data having a PSPI of 4, and that only the LFA-CRNN was able to detect data having a PSPI of 8. To sum up all experiments, the proposed LFA-CRNN model showed a stable graph in the learning process, and in the performance evaluation, using the test data, it showed the highest performance of 98.92%. In addition, its loss measurement showed the lowest result at approximately 0.036. Although the LFA-CRNN's precision and recall were quite poor, its average precision was 75% (as high as the precision by AlexNet), and it showed the highest average recall at 75%.

The LFA-CRNN proposed in this study showed higher accuracy, using fewer input dimensions than comparable models. We judge that this is because of the effect of the maximum removal of unnecessary regions. We examined the metadata necessary for analyzing test data of facial expressions and judged that the color and area (size) constituting the images were unnecessary elements. Thus, the remaining element was the information about segments, and we set up a hypothesis for a sentiment analysis algorithm through this. When people analyze facial expressions, they do not usually consider colors, and the color element was removed, using the understanding of emotions through the shapes of the mouth, eyes, and eyebrows. Also, the images with colors removed were similar to the images expressed with the outline. In learning with the neural network model, a big loss of data took place when images were reduced via max-pooling and stride in the processing, and the overfitting and wind-up phenomena occurred. Thus, we devised a method for reducing the size of the images, and that method is LFA. LFA maintained information about segments as much as possible to prevent data loss that might occur during processing, utilizing data with both color and unnecessary areas removed. In other words, when we extracted emotions, necessary elements were maintained as much as possible, and all other information was minimized. We judge that LFA-CRNN shows high accuracy for these reasons.

5. Conclusions

With this paper's proposed method, health risks due to an abnormal health status that may occur while someone is driving are determined through facial expressions, a representative medium capable of confirming a person's emotional state based on external clues. The purpose of this study was to construct a system capable of preventing traffic accidents and secondary accidents resulting from chronic diseases, which are increasing as our society ages. Although automated driving systems are being mounted on vehicles and are commercialized based on vehicle technology advancements, such systems do not take into consideration driver status. If abnormal health status in a driver is detected while the vehicle is in motion, it may operate normally, but the drivers might not be able to meet the required "golden time" to address any health problem that arises. Our system checks the driver's health status based on facial expressions in order to resolve to a certain extent problems related to chronic diseases. To do so, in this paper, an LFA dimensionality reduction algorithm was used to reduce the size of input images, and the LFA-CRNN model receiving the reduced LFA data as input was designed and used to classify the status of drivers as being in pain or not. The LFA is a method where a series of filters is used to assign a unique number to the line-segment information that makes up a facial image, and then, the input image is converted into a two-dimensional array having a size of

16 × 16 by adding up the unique numbers. As the converted data are learned through the LFA-CRNN model, facial expressions indicating pain are classified. To evaluate performance, a comparison was made with pre-existing CRNN and AlexNet models. The UNBC-McMaster database was used to learn pain-related expressions. As far as the accuracy and loss calculated through learning are concerned, the LFA-CRNN showed the highest accuracy at 98.92%, a CRNN alone showed accuracy of 98.21%, and AlexNet showed accuracy of 97.4%. In addition, the LFA-CRNN showed the lowest loss at approximately 0.036, the CRNN showed a loss of 0.045, and AlexNet showed a loss of 0.117. Although the LFA-CRNN's precision and recall measurement results were quite poor, average precision was 75%, which is as high as the 75% precision achieved by AlexNet.

We optimized the facial expressions and the data sources for the LFA-CRNN, and intend to compare the processing times of several models and improve the accuracy in the future. The proposed LFA-CRNN algorithm shows high dependency on the outline detection method. This is self-evident, because LFA is based on segment analysis. We are devising an outline detection technique that can optimally be applied to LFA based on this fact. In addition, the LFA performance process generates a one-dimensional sequence before the production of a two-dimensional LS-Map. It is expected that by converting this, a class can be produced that can be used in the neural network model. Through this improvement process, we will combine the LFA-CRNN model with a system for recognition of facial expressions and motions that can be used in services like smart homes and smart health care, and we plan to apply that to mobile edge computing systems and video security.

Author Contributions: K.C. and R.C.P. conceived and designed the framework. E.J.H. and C.-M.K. implemented LFA-CRNN model. R.C.P. and C.-M.K. performed experiments and analyzed the results. All authors have contributed in writing and proofreading the paper. All authors have read and agreed to the published version of the manuscript.

Funding: This work is supported by the Korea Agency for Infrastructure Technology Advancement (KAIA) grant funded by the Ministry of Land, Infrastructure and Transport (Grant 20CTAP-C157011-01).

Conflicts of Interest: The authors declare no conflict of interest.

References

1. Yeem, M.J.; Song, H.J. The Effect of facial emotion Recognition of Real-face Expression and Emoticons on Interpersonal Competence: Mobile Application Based research for Middle School Students. *J. Emot. Behav. Disord.* **2019**, *35*, 265–284.
2. Olderbak, S.G.; Wilhelm, O.; Hildebrandt, A.; Quoidbach, J. Sex differences in facial emotion perception ability across the lifespan. *Cogn. Emot.* **2018**, *33*, 579–588. [CrossRef]
3. Poria, S.; Majumder, N.; Mihalcea, R.; Hovy, E.; Majumderd, N.; Mihalceae, R. Emotion Recognition in Conversation: Research Challenges, Datasets, and Recent Advances. *IEEE Access* **2019**, *7*, 100943–100953. [CrossRef]
4. Kang, X.; Ren, F.; Wu, Y. Exploring Latent Semantic Information for Textual Emotion Recognition in Blog Articles IEEE/CAA. *J. Autom. Sin.* **2018**, *5*, 204–216.
5. Guo, J.; Lei, Z.; Wan, J.; Avots, E.; Hajarolasvadi, N.; Knyazev, B.; Kuharenko, A.; Junior, J.C.S.J.; Baró, X.; Demirel, H.; et al. Dominant and Complementary Emotion Recognition from Still Images of Faces. *IEEE Access* **2018**, *6*, 26391–26403. [CrossRef]
6. Perlovsky, L.; Schoeller, F. Unconscious emotions of human learning. *Phys. Life Rev.* **2019**, *31*, 257–262. [CrossRef]
7. Chung, K.; Park, R.C. P2P-based open health cloud for medicine management. *Peer-to-Peer Netw. Appl.* **2019**, *13*, 610–622. [CrossRef]
8. Kim, J.; Jang, H.; Kim, J.T.; Pan, H.-J.; Park, R.C. Big-Data Based Real-Time Interactive Growth Management System in Wireless Communications. *Wirel. Pers. Commun.* **2018**, *105*, 655–671. [CrossRef]
9. Kim, J.-C.; Chung, K. Prediction Model of User Physical Activity using Data Characteristics-based Long Short-term Memory Recurrent Neural Networks. *KSII Trans. Internet Inf. Syst.* **2019**, *13*, 2060–2077. [CrossRef]

10. Baek, J.-W.; Chung, K. Context Deep Neural Network Model for Predicting Depression Risk Using Multiple Regression. *IEEE Access* **2020**, *8*, 18171–18181. [CrossRef]
11. Baek, J.-W.; Chung, K. Multimedia recommendation using Word2Vec-based social relationship mining. *Multimed. Tools Appl.* **2020**, 1–17. [CrossRef]
12. Kang, J.-S.; Shin, D.H.; Baek, J.-W.; Chung, K. Activity Recommendation Model Using Rank Correlation for Chronic Stress Management. *Appl. Sci.* **2019**, *9*, 4284. [CrossRef]
13. Chung, K.; Kim, J. Activity-based nutrition management model for healthcare using similar group analysis. *Technol. Health Care* **2019**, *27*, 473–485. [CrossRef]
14. Haz, H.; Ahuja, S. Latest trends in emotion recognition methods: Case study on emotiw challenge. *Adv. Comput. Res.* **2020**, *10*, 34–50. [CrossRef]
15. Song, X.; Chen, Y.; Feng, Z.-H.; Hu, G.; Zhang, T.; Wu, X.-J. Collaborative representation based face classification exploiting block weighted LBP and analysis dictionary learning. *Pattern Recognit.* **2019**, *88*, 127–138. [CrossRef]
16. Nassih, B.; Amine, A.; Ngadi, M.; Hmina, N. DCT and HOG Feature Sets Combined with BPNN for Efficient Face Classification. *Procedia Comput. Sci.* **2019**, *148*, 116–125. [CrossRef]
17. Lenc, L.; Kral, P. Automatic face recognition system based on the SIFT features. *Comput. Electr. Eng.* **2015**, *46*, 256–272. [CrossRef]
18. Taigman, Y.; Yang, M.; Ranzato, M.; Wolf, L. DeepFace: Closing the Gap to Human-Level Performance in Face Verification. In Proceedings of the 2014 IEEE Conference on Computer Vision and Pattern Recognition, Columbus, OH, USA, 23–28 June 2014; Institute of Electrical and Electronics Engineers (IEEE): Piscataway, NJ, USA, 2014; pp. 1701–1708.
19. Schroff, F.; Kalenichenko, D.; Philbin, J. FaceNet: A unified embedding for face recognition and clustering. In Proceedings of the 2015 IEEE Conference on Computer Vision and Pattern Recognition (CVPR), Boston, MA, USA, 7–12 June 2015; pp. 815–823.
20. Luttrell, J.; Zhou, Z.; Zhang, C.; Gong, P.; Zhang, Y.; Iv, J.B.L. Facial Recognition via Transfer Learning: Fine-Tuning Keras_vggface. In Proceedings of the 2017 International Conference on Computational Science and Computational Intelligence (CSCI), Las Vegas, NV, USA, 14–16 December 2017; pp. 576–579.
21. Sun, Y.; Wang, X.; Tang, X. Deep Learning Face Representation by Joint Identification-Verification. *arXiv* **2014**, arXiv:1406.4773.
22. Sun, Y.; Liang, D.; Wang, X.; Tang, X. DeepID3: Face Recognition with Very Deep Neural Networks. In Proceedings of the 2015 IEEE Conference on Computer Vision and Pattern Recognition (CVPR), Boston, MA, USA, 7–12 June 2015.
23. Khan, R.U.; Zhang, X.; Kumar, R. Analysis of ResNet and GoogleNet models for malware detection. *J. Comput. Virol. Hacking Tech.* **2018**, *15*, 29–37. [CrossRef]
24. Muhammad, G.; Alsulaiman, M.; Amin, S.U.; Ghoneim, A.; Alhamid, M.F. A Facial-Expression Monitoring System for Improved Healthcare in Smart Cities. *IEEE Access* **2017**, *5*, 10871–10881. [CrossRef]
25. Lim, K.-T.; Won, C. Face Image Analysis using Adaboost Learning and Non-Square Differential LBP. *J. Korea Multimed. Soc.* **2016**, *19*, 1014–1023. [CrossRef]
26. Kang, H.; Lim, K.-T.; Won, C. Learning Directional LBP Features and Discriminative Feature Regions for Facial Expression Recognition. *J. Korea Multimed. Soc.* **2017**, *20*, 748–757. [CrossRef]
27. Jabon, M.E.; Bailenson, J.N.; Pontikakis, E.; Takayama, L.; Nass, C. Facial expression analysis for predicting unsafe driving behavior. *IEEE Pervasive Comput.* **2010**, *10*, 84–95. [CrossRef]
28. Agbolade, O.; Nazri, A.; Yaakob, R.; Ghani, A.A.; Cheah, Y.K. 3-Dimensional facial expression recognition in human using multi-points warping. *BMC Bioinform.* **2019**, *20*, 619. [CrossRef] [PubMed]
29. Park, B.-H.; Oh, S.-Y.; Kim, I.-J. Face alignment using a deep neural network with local feature learning and recurrent regression. *Expert Syst. Appl.* **2017**, *89*, 66–80. [CrossRef]
30. Wang, Y.; Li, Y.; Song, Y.; Rong, X. Facial Expression Recognition Based on Random Forest and Convolutional Neural Network. *Informatics* **2019**, *10*, 375. [CrossRef]
31. Jeong, M.; Ko, B.C. Driver's Facial Expression Recognition in Real-Time for Safe Driving. *Sensors* **2018**, *18*, 4270. [CrossRef]
32. Ra, S.T.; Kim, H.J.; Lee, S.H. A Study on Deep Learning Structure of Multi-Block Method for Improving Face Recognition. *Inst. Korean Electr. Electron. Eng.* **2018**, *22*, 933–940.
33. Facereader. Available online: https://www.noldus.com/facereader/ (accessed on 16 December 2019).

34. Neighbor System of Korea. Available online: http://www.neighbor21.co.kr/ (accessed on 3 January 2020).
35. Chung, K.; Shin, D.H.; Park, R.C. Detection of Emotion Using Multi-Block Deep Learning in a Self-Management Interview App. *Appl. Sci.* **2019**, *9*, 4830. [CrossRef]
36. Yuan, Q.; Xiao, N. Scaling-Based Weight Normalization for Deep Neural Networks. *IEEE Access* **2019**, *7*, 7286–7295. [CrossRef]
37. Pan, S.; Zhang, W.; Zhang, W.; Xu, L.; Fan, G.; Gong, J.; Zhang, B.; Gu, H. Diagnostic Model of Coronary Microvascular Disease Combined with Full Convolution Deep Network with Balanced Cross-Entropy Cost Function. *IEEE Access* **2019**, *7*, 177997–178006. [CrossRef]
38. Zhang, S.; Wang, Y.; Liu, M.; Bao, Z. Data-Based Line Trip Fault Prediction in Power Systems Using LSTM Networks and SVM. *IEEE Access* **2017**, *6*, 7675–7686. [CrossRef]
39. Hu, Y.; Jin, Z.; Wang, Y. State Fusion Estimation for Networked Stochastic Hybrid Systems with Asynchronous Sensors and Multiple Packet Dropouts. *IEEE Access* **2018**, *6*, 10402–10409. [CrossRef]
40. Liu, L.; Luo, Y.; Shen, X.; Sun, M.; Li, B. β-Dropout: A Unified Dropout. *IEEE Access* **2019**, *7*, 36140–36153. [CrossRef]
41. Peng, D.; Liu, Z.; Wang, H.; Qin, Y.; Jia, L. A Novel Deeper One-Dimensional CNN with Residual Learning for Fault Diagnosis of Wheelset Bearings in High-Speed Trains. *IEEE Access* **2019**, *7*, 10278–10293. [CrossRef]
42. Shi, B.; Bai, X.; Yao, C. An End-to-End Trainable Neural Network for Image-Based Sequence Recognition and Its Application to Scene Text Recognition. *IEEE Trans. Pattern Anal. Mach. Intell.* **2017**, *39*, 2298–2304. [CrossRef]
43. Han, X.; Zhong, Y.; Cao, L.; Zhang, L. Pre-Trained AlexNet Architecture with Pyramid Pooling and Supervision for High Spatial Resolution Remote Sensing Image Scene Classification. *Remote. Sens.* **2017**, *9*, 848. [CrossRef]
44. Lucey, P.; Cohn, J.F.; Prkachin, K.M.; Solomon, P.E.; Matthews, I. Painful data: The UNBC-McMaster shoulder pain expression archive database. *Face Gesture* **2011**, 57–64. [CrossRef]
45. Fawcett, T. An introduction to ROC analysis. *Pattern Recognit. Lett.* **2006**, *27*, 861–874. [CrossRef]

applied sciences

MDPI

Article

Application of Texture Descriptors to Facial Emotion Recognition in Infants

Ana Martínez †, Francisco A. Pujol *,† and Higinio Mora †

Department of Computer Technology, University of Alicante, 03690 San Vicente del Raspeig-Alicante, Spain; amlopez.ana@gmail.com (A.M.); hmora@ua.es (H.M.)

* Correspondence: fpujol@ua.es

† These authors contributed equally to this work.

Received: 3 December 2019; Accepted: 30 January 2020; Published: 7 February 2020

check for updates

Featured Application: A system to detect pain in infants using facial expressions has been developed. Our system can be easily adapted to a mobile app or a wearable device. The recognition rate is above 95% when using the Radon Barcodes (RBC) descriptor. It is the first time that RBC is used in facial emotion recognition.

Abstract: The recognition of facial emotions is an important issue in computer vision and artificial intelligence due to its important academic and commercial potential. If we focus on the health sector, the ability to detect and control patients' emotions, mainly pain, is a fundamental objective within any medical service. Nowadays, the evaluation of pain in patients depends mainly on the continuous monitoring of the medical staff when the patient is unable to express verbally his/her experience of pain, as is the case of patients under sedation or babies. Therefore, it is necessary to provide alternative methods for its evaluation and detection. Facial expressions can be considered as a valid indicator of a person's degree of pain. Consequently, this paper presents a monitoring system for babies that uses an automatic pain detection system by means of image analysis. This system could be accessed through wearable or mobile devices. To do this, this paper makes use of three different texture descriptors for pain detection: Local Binary Patterns, Local Ternary Patterns, and Radon Barcodes. These descriptors are used together with Support Vector Machines (SVM) for their classification. The experimental results show that the proposed features give a very promising classification accuracy of around 95% for the Infant COPE database, which proves the validity of the proposed method.

Keywords: emotion recognition; pattern recognition; texture descriptors; mobile tool

1. Introduction

Facial expressions are one of the most important stimuli when interpreting social interaction, as they provide information on the identity of the person and on his emotional state. Facial emotions are one of the most important signal systems when expressing to other people what happens to human beings [1].

The recognition of facial expressions is especially interesting because it allows for detecting feelings and moods in people, which are applicable in fields such as psychology, teaching, marketing or even health, which is the main objective of this work.

The automatic recognition of facial expressions could be a great advance in the field of health, in applications such as pain detection in people unable to communicate verbally, decreasing the continuous monitoring by medical staff, or for people with Autism Spectrum Disorder, for instance, who have difficulty when understanding other people emotions.

Babies are one of the biggest groups that cannot express pain verbally, so this impossibility has created the necessity of using other media for its evaluation and detection. In this way, pain scales based on vital signals and facial changes have been created to evaluate the pain of neonates [2]. Thus, the main objective of this paper is to create a tool which reduces the continuous monitoring by parents and medical staff. For that purpose, a set of computer vision methods with supervised learning have been implemented, making it feasible to develop a mobile application to be used in a wearable device. For the implementation, this paper has used the Infant COPE database [3], a database composed of 195 images of neonates, which is one of the few available public databases for infants' pain detection.

Regarding pain detection using computer vision, several previous studies have been carried out. Thus, Roy et al. [4] proposed the extraction of facial features for automatic pain detection in adults, using the NBC-McMaster Shoulder Pain Expression Archive Database [5,6]. Using the same database, Lucey et al. [7] developed a system that classifies pain in adults after extracting facial action units. More recently, Rodriguez et al. [8] used Convolutional Neural Networks (CNNs) to recognize pain from facial expressions and Ilyas et al. [9] implemented a facial expression recognition system for traumatic brain injured patients. However, when focusing on detecting pain in babies, very few works can be found. Among them, Brahnam et al. used Principal Components Analysis (PCA) reduction for feature extraction and Support Vector Machines (SVM) for classification in [10], obtaining a recognition rate of up to 88% using a grade 3 polynomial kernel. Then, in [11], Mansor and Rejab used Local Binary Patterns (LBP) for the extraction of characteristics, while, for classification, Gaussian and Nearest Mean Classifier were used. With these tools, they achieved a success rate of 87.74–88% for the Gaussian Classifier and of 76–80% with the Nearest Mean Classifier. Similarly, Local Binary Patterns were used as well in [12] for feature extraction and SVM for classification, obtaining an accuracy of 82.6%. More recently, and introducing deep learning methods, Ref. [13] fused LBP, Histogram of Oriented Gradients (HOG), and CNNs as feature extractors, with SVM for classification, with an accuracy of 83.78% as the best result. Then, in [14], Zamzmi et al. used pre-trained CNNs and a strain-based expression segmentation algorithm as a feature extractor together with a Naive Bayes (NB) classifier, obtaining a recognition accuracy of 92.71%. In [15], Zamzmi et al. proposed an end-to-end Neonatal Convolutional Neural Network (N-CNN) for automatic recognition of neonatal pain, obtaining an accuracy of 84.5%. These works validated their proposed methods using the Infant COPE database mentioned above.

Other recent works tested their proposed methods with other databases. Thus, an automatic discomfort detection system for infants by analyzing their facial expressions in videos from a dataset collected at the hospital Maxima Medical Center in Veldhoven, The Netherlands, was presented in [16]. The authors used again HOG, LBP and SVM with 83.1% correctly detected discomfort expressions. Finally, Zamzmi et al. [14] used CNNs with transfer learning as a pain expression detector, achieving 90.34% accuracy in a dataset recorded at Tampa General Hospital and in [15] obtained an accuracy of 91% for the NPAD database.

On the other hand, concerning emotion recognition and wearable devices, most of the proposed methods until now relied on biomedical signals [17–20]. When using images, and more specifically, facial images to recognize emotions, very few wearable devices can be found. Among them, one can find the work by Kwon et al. [21], where they proposed a glasses-type wearable system to detect a user's emotion using facial expression and physiological responses, reaching around 70% in the subject-independent case and 98% in the subject-dependent one. In [22], another system to automate facial expression recognition that runs on wearable glasses is proposed, reaching a 97% classification accuracy for eight different emotions. More recently, Kwon and Kim described in [23] another glassed-type wearable device to detect emotions from a human face via multi-channel facial responses, obtaining an accuracy of 78% at classifying emotions. Wearable devices have also been designed for infants to monitor vital signs [24], body temperature [25], health using an electrocardiogram (ECG) sensor [26], or as a pediatric rehabilitation device [27]. In addition, there is a growing number of mobile applications for infants, such as SmartCED [28], which is an Android application for epilepsy diagnosis, or commercial devices with a smartphone application for parents [29] (smart socks [30] or the popular video monitors [31]).

However, no smartphone application or wearable device related to pain detection through facial expression recognition in infants has been found. Therefore, this work investigates different methods to implement a reliable tool to assist in the automatic detection of pain in infants using computer vision and supervised learning, extending our previous work presented in [2]. As mentioned before, texture descriptors and, specifically, Local Binary Patterns, are among the most popular algorithms to extract features for facial emotions recognition. Thus, this work will compare the results after applying several texture descriptors, including Radon Barcodes, which is the first time that they are used to detect facial emotions, this being the main contribution of this paper. Moreover, our tool can be easily implemented in a wireless and wearable system, so it could have many potential applications, such as alerting parents or medical staff quickly and efficiently when a baby is in pain.

This paper is organized as follows: Section 2 explains the main features about the methods used in our research and outlines the proposed method; Section 3 describes the experimental setup and the set of experiments completed and their interpretation; and, finally, conclusions and some future works are discussed in Section 4.

2. Materials and Methods

In this section, some theoretical concepts are explained first. Then, at the end of the section, the method followed to determine whether a baby is in pain or not is described.

2.1. Pain Perception in Babies

Traditionally, babies' pain has been undervalued, receiving limited attention due to the thought of babies suffering less pain than adults because of their supposed 'neurological immaturity' [32,33]. This has been refuted through several studies over the last few years, especially by the one conducted by the John Radcliffe Hospital in Oxford in 2015 [34], which concluded that infants' brains react in a very similar way to adult brains when they are exposed to the same pain stimulus. Recent works suggest that infants' units in hospitals must adopt reliable pain assessment tools, since they may derive in short- and long-term sequels [35,36].

As mentioned before, the impossibility of expressing pain in a verbal way has created the need of using other media to assess pain, detect it, and take the appropriate actions. This is why pain assessment scales based on behavioral indicators has been created, such as PIPP (Premature Infant Pain Profile) [37], CRIES (Crying; Requires increased oxygen administration; Increased vital signs; Expression; Sleeplessness) [38], NIPS (Neonatal Infant Pain Scale) [39], or NFCS (Neonatal Facial Coding System) [40,41]. While most assessment scales use vital signals such as heart rate or oxygen saturation, NFCS is based on facial changes through face muscles, mainly on forehead protrusion, contraction of eyelids, nasolabial groove, horizontal stretch of the mouth, and tense tongue [42]. Figure 1 shows a graphical example of the NFCS scale. As this paper uses an image database, this last scale is ideal to determine if the babies are or not in pain, by analyzing the facial changes in different areas according to the NFCS scale.

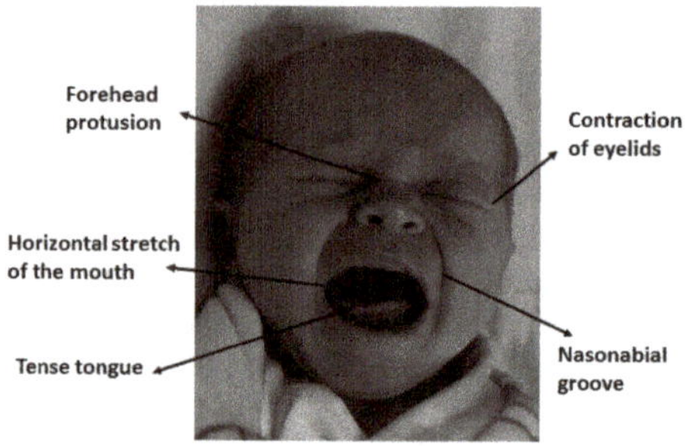

Figure 1. Facial expression of physical distress is the most consistent behavioral indicator of pain in infants.

2.2. Feature Extraction

Feature extraction methods of facial expressions can be divided depending on their approach. Generally speaking, features are extracted from facial deformation, which is characterized by changes in shape and texture, and from facial motion, which is characterized by either the speed and direction of movement or deformations in the face image [43,44].

As explained in the last section, in this paper, the NFCS scale has been selected, since its reliability, validity, and clinical utility has been extensively proved [45,46]. The criteria of classification of pain in the NFCS scale is based on facial deformations and it depends on the texture of the face. Texture descriptors have been widely used in machine learning and pattern recognition, being successfully applied to object detection, face recognition, and facial expression analysis, among other applications [47]. Consequently, three texture descriptors are taken into account in this research: the popular Local Binary Pattern descriptor; then, a variation of this descriptor, the Local Ternary Patterns; and, finally, a recently proposed descriptor, the Radon Barcodes, which are based on the Radon transform.

2.2.1. Local Binary Patterns

Local Binary Patterns (LBP) are a simple but effective texture descriptor which label every pixel of the image analyzing its neighborhood. It identifies if the grey level of every neighbor pixel is above a certain threshold and codifies this comparison with a binary number. This descriptor has become very popular due to its good classification accuracy and its low computational cost, which allows real-time image processing in many applications. In addition, this descriptor has a great robustness when there are varying lighting conditions [48,49].

On its basic version, LBP operator works with a 3×3 matrix that goes across the image pixel by pixel, identifying the grey values of its eight neighbors and taking as a threshold the grey value of the central pixel. Thus, the binary code is obtained as follows: if the neighbor pixels has a lower value than the central one, they will coded as 0; otherwise, their code will be 1. Finally, each binary value is weighted by its corresponding power of two and added to obtain the LBP code of the pixel. In Figure 2, a graphic example is shown.

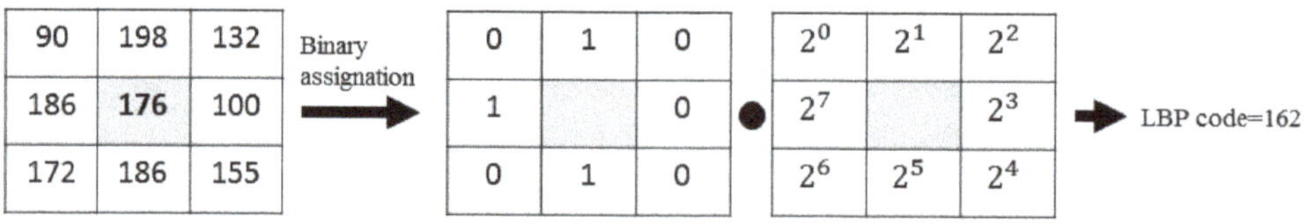

Figure 2. Graphic example of the LBP descriptor.

This descriptor has been extended over the years, so that it can be used in circle neighborhoods of different sizes. In this circular version, neighbors are equally spaced, allowing the use of any radio and any number of neighboring pixels. Once the codes of all pixels are obtained, a histogram is created. It is also common to divide the image into cells, so that a histogram per cell would be obtained, being finally concatenated. In addition, the LBP descriptor has uniformity, which reduces negligible information significantly, and therefore it provides low computational cost and invariance to rotations, which become two important properties when applied to facial expression recognition in mobile and wearable devices [50].

2.2.2. Local Ternary Patterns

Tan and Triggs [51] presented a new texture operator which is more robust to noise than LBP in uniform regions. It consists of an LBP extended into 3-valued codes (0, 1, −1). Figure 3 shows a practical example of how Local Ternary Patterns (LTP) work: first, threshold t is established. Then, if any neighbor pixel has a value below the value of the central pixel minus the threshold, it is assigned −1 and, if the value is over the value of the central pixel plus the threshold, it is assigned 1. Otherwise, it is assigned 0. After the thresholding step, the upper pattern and lower pattern are constructed as follows: for the upper pattern, all 1's are assigned 1, and the rest of the values (0s and −1's) are assigned 0; for the lower pattern, all −1's are assigned 1, and the rest of the values (0s and 1's) are assigned 0. Finally, both patterns are encoded in two different binary codes, so this descriptor provides two binary codes for one pixel instead of one as LBP does, that is, more information about the texture of the image. All of this process is shown in Figure 3.

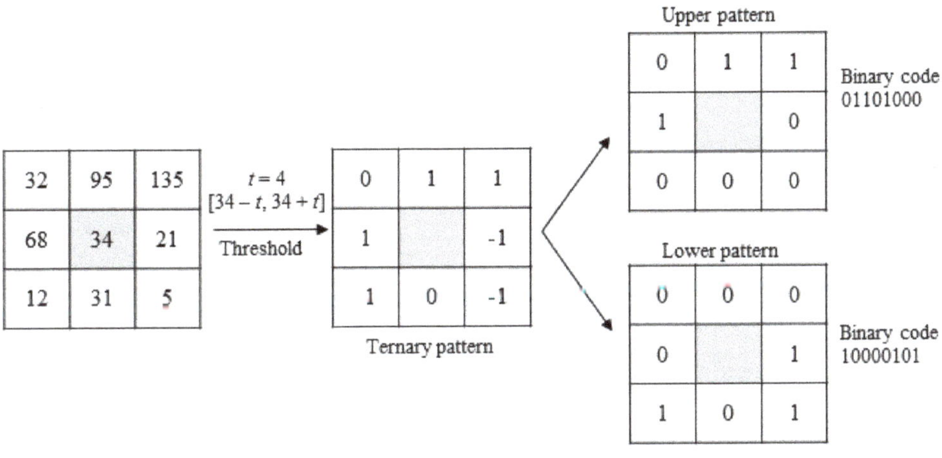

Figure 3. Graphic example of the LTP descriptor.

The LTP operator has been applied successfully to similar applications as LBP, including medical images, human action classification and facial expression recognition, among others.

2.2.3. Radon Barcodes

The Radon Barcodes (RBC) operator is based on the Radon transform, which is having an increasing interest in image processing, since it is extremely robust to noise and presents scale and rotation invariance [52,53]. Moreover, it has been used for years to process medical images, and is the basis of current computerized tomography. As mentioned before, facial expression features are based on facial deformations and involve changes in shape, texture, and motion. As Radon transform presents valuable features regarding image translation, scaling, and rotation, its application to facial recognition of emotions has been considered in this work.

Essentially, Radon transform consists of an integral transform which projects all pixels from different orientations to a single vector. Consequently, RBCs are basically the sum (integral) of the values along lines constituted by different angles. Thus, Radon transform is first applied to any input image, and then projections are performed. Finally, all the projections are thresholded individually to generate code sections, which are concatenated to build the Radon Barcode. A simple way for thresholding the projection is to calculate a typical value using the median operator applied on all non-zero values of each projection [53]. Algorithm 1 shows how RBC works [53] and in Figure 4 a graphic example is shown.

Algorithm 1: Radon Barcode Generation [53]

Initialize Radon Barcode $r \leftarrow \emptyset$
Initialize angle $\theta \leftarrow$ and $R_N = C_N \leftarrow 32$
Normalize the input image $\bar{I} = \text{Normalize}(I, R_N, C_N)$
Set the number of projection angles, e.g., $n_p \leftarrow 8$
while $\theta < 180$ **do**
  Get all projections p for θ
  Find typical value $T_{typical} \leftarrow median_i(p_i)|_{p_i \neq 0}$
  Binarize projections: $b \leftarrow p \geq T_{typical}$
  Append the new row $r \leftarrow$ append(r, b)
  $\theta \leftarrow \theta + \frac{180}{n_p}$
end
Return r

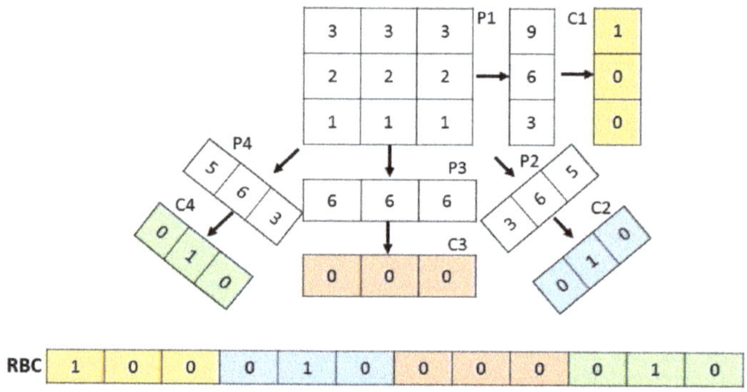

Figure 4. Graphic example of an RBC descriptor.

Until now, the main application of Radon Barcodes comes from medical image retrieval, where it has given high accuracy. As in the recognition of facial expressions robustness in orientation, illumination, and scale changes are needed, we consider that the RBC descriptor can be a good technique to provide a reliable classification of pain/non-pain in infants using facial images, being the first time that RBC are used in these kinds of applications.

2.3. Classification: Support Vector Machines

In order to classify properly the features extracted using any of the descriptors defined above, Support Vector Machines (SVM) are chosen.

The main idea of SVM is to select a hyperplane that is equidistant to the training examples of every class to be classified so that the so-called maximum margin hyperplane between classes is obtained [54,55]. To define this hyperplane, only the training data of each class that fall right next to those margins are taken into account, which are called support vectors. In this work, this hyperplane would be the one which separates the characteristics obtained from pain and non-pain facial images. In cases where a linear function does not allow for separating the examples properly, a nonlinear SVM is used. To define the hyperplane in this case, the input space of the examples $\mathbb{X}$ is transformed into a new one, $\Phi(\mathbb{X})$, where a linear separation hyperplane is constructed using kernel functions as they are represented in Figure 5. A kernel function $K(x, x')$ is a function that assigns to each pair of elements $x, x' \in \mathbb{X}$ a real value corresponding to the scalar product of the transformed version of that element in a new space. There are several types of kernel, such as:

- Linear kernel:

$$K\left(x, x'\right) = < x, x' >, \tag{1}$$

- P-Grade polynomial kernel:

$$K\left(x, x'\right) = \left[\gamma < x, x' > + \tau\right]^{p}, \tag{2}$$

- Gaussian kernel:

$$K\left(x, x'\right) = \left[exp(-\gamma \left\|x, x'\right\|^{2})\right], \qquad \gamma > 0, \tag{3}$$

where $\gamma > 0$ is a scaling parameter and τ is a constant.

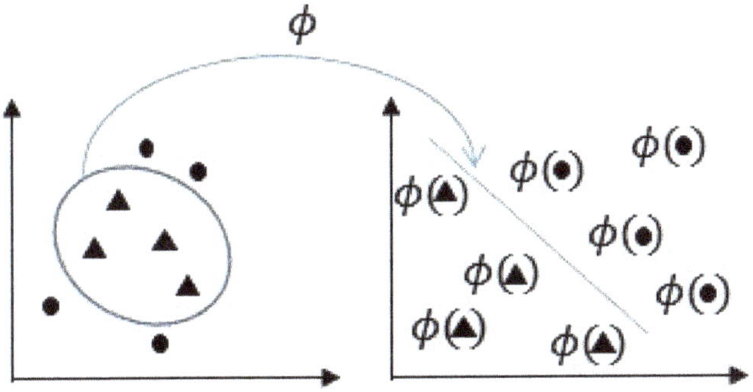

Figure 5. Representation of the transformed space for nonlinear SVM.

The selection of the kernel depends on the application and situation, and a linear kernel is recommended when the linear separation of data is simple. In the rest of the cases, it will be necessary to experiment with the different functions to obtain the best model for each case, since kernels use different algorithms and parameters.

Once the hyperplane is obtained, it will be transformed back into the original space, thus obtaining a nonlinear decision boundary [2].

2.4. The Proposed Method

Our application has been implemented in MATLAB© R2017. The toolboxes that have been used are *Statistics and Machine Learning* and *Computer Vision System*. As mentioned in Section 1, for the development of the tool, the Infant COPE database [3] has been used. This is a database that is composed of 195 color images of 26 neonates, 13 boys, and 13 girls, with an age between 18 hours and 3 days. For the images, the neonates have been exposed to the pain of the heel test and to three non-painful stimuli: a corporal disturbance (movement from one cradle to another), air stimulation applied to the nose, and the friction of a wet cotton in the heel. In addition, images of resting infants have been taken.

As mentioned before, this implementation could be applied to a mobile device and/or a wearable system, so that, on the one hand, a baby monitor would continuously analyze the images it captures. On the other hand, the parents or medical staff would wear a bracelet or have a mobile application to warn them when the baby is suffering pain. The diagram in Figure 6 shows a possible example of the implementation stages.

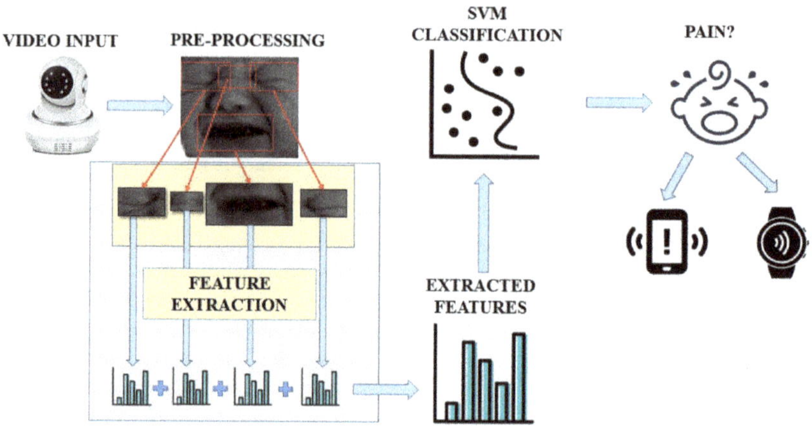

Figure 6. Flowchart of the different stages.

The first step is pre-processing the input image by detecting infants' faces and then resizing the resulting images and converting them into grey scale. All images are normalized to a size of 100 × 120 pixels. Afterwards, features have been extracted using the texture descriptors mentioned before. The NFCS scale will be followed, so descriptors have been applied only to relevant facial areas to the NFCS scale: right eye, left eye, mouth, and brow. These areas are manually selected with sizes 30 × 50 pixels for eyes, 40 × 90 pixels for mouth, and 15 × 40 pixels for brow. It was possible to make an analysis to find the ideal sizes for each part due to the small size of the used database. Feature vectors from each area have been concatenated to obtain the global descriptor.

Finally, a previously trained SVM classifier decides if the input frame corresponds with a baby in pain or not. The system will be continuously monitoring the video frames obtained and sending an alarm to the mobile device if a pain expression is detected.

3. Results

In this section, a comparison of three different methods for feature extraction is completed: Local Binary Patterns, Local Ternary Patterns, and Radon Barcodes. According to the results obtained in [2], a Gaussian Kernel has been chosen for SVM classification, since it provides an optimal behavior for the Infant COPE database. SVM has been trained with 13 pain images and 13 non-pain images, and the tests have been performed with 30 pain images and 93 non-pain images different from the training stage. The unbalanced number of images is due to the number of pictures of each class available in the database.

To evaluate the tests, confusion matrices, cross-validation and error rate have been used. In this case, error rate has been calculated as the number of incorrect predictions divided by the total number of evaluated predictions.

3.1. Results on LBP

The parameters to be considered on the LBP descriptor are the radius, the number of neighbors and the cell sizes. As mentioned before, images has been previously cropped into four different areas. According to the previous results in [2], the best recognition rate is obtained when each of these areas is not divided into cells. Therefore, as it is shown in Figure 7, the recognition rates for all the possible combinations with radius 1, 2, and 3, and neighbors 8, 10, 12, 16, 18, 20, and 24 have been calculated to select the optimum values.

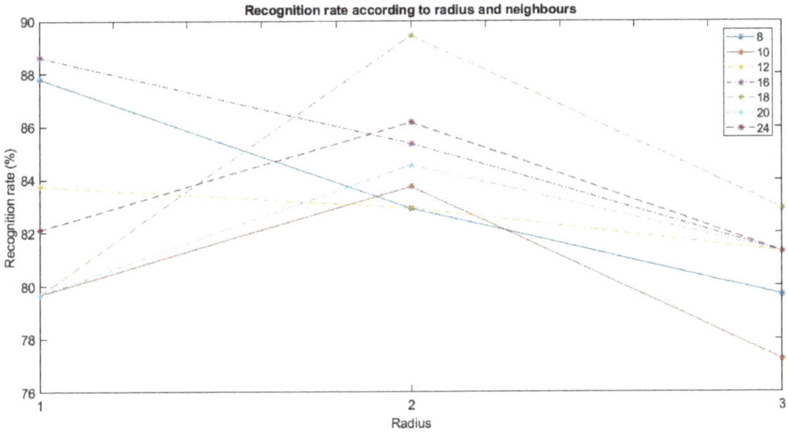

Figure 7. Recognition rate according to radius and neighbors.

As shown in Figure 7, the parameters with the best recognition rate are radius 2 and 18 neighbors. This combination presents the following confusion matrix CM_{LBP}:

$$CM_{LBP} = \begin{pmatrix} 27 & 3 \\ 10 & 83 \end{pmatrix} \tag{4}$$

It implies that there are three false positives and 10 false negatives, thus having an error rate of 10.57% and, therefore, a successful recognition rate of 89.43%.

3.2. Results on LTP

In this case, the parameters to be calculated on the LTP descriptor are the same as in LBP, but adding threshold t. Let us consider the same values for the parameters which gave the best result for LBP (radius 2 and 18 neighbors), and values from $t = 1$ to 10 for the threshold have been chosen.

As is shown in Figure 8, the best result is obtained for threshold $t = 6$, which presents the next confusion matrix CM_{LTP}:

$$CM_{LTP} = \begin{pmatrix} 20 & 10 \\ 3 & 90 \end{pmatrix} \tag{5}$$

It implies that there are 10 false positives and three false negatives, thus having an error rate of 10.57% and, therefore, a recognition rate of 89.43%.

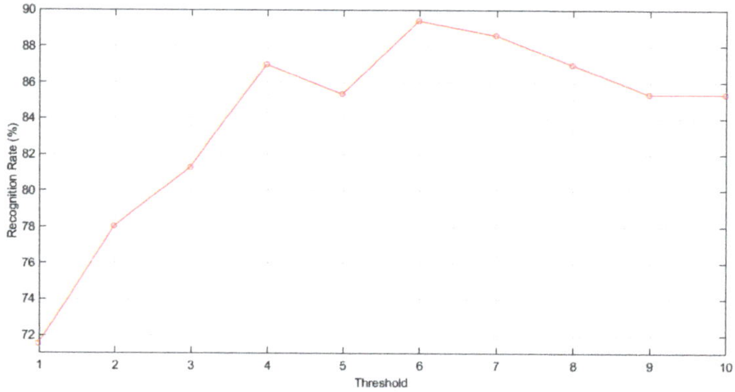

Figure 8. Recognition rate according to LTP threshold.

3.3. Results on RBC

The parameter to be calculated in the RBC method is the number of projection angles. To do this, typical values 4, 8, 16, and 32, as considered in [53], have been chosen. The results of the carried tests are shown in Figure 9.

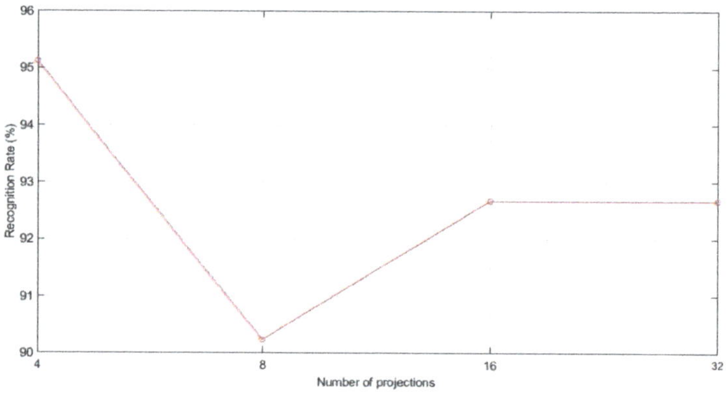

Figure 9. Recognition rate according to RBC projections.

As we can see in Figure 9, the best result is obtained with four projections, which presents the next confusion matrix CM_{RBC}:

$$CM_{RBC} = \begin{pmatrix} 27 & 3 \\ 3 & 90 \end{pmatrix} \quad (6)$$

It implies that there are three false positives and three false negatives, thus having an error rate of 4.88% and, therefore, a recognition rate of 95.12%.

3.4. Final Results and Discussion

As shown throughout this section, the best results are obtained by RBC with a recognition rate of 95.12%, followed by LBP and LTP with a recognition rate of 89.43 %. These results show the validity of applying Radon Barcodes to facial emotion recognition, as seen in Section 2, and it can be then concluded that the RBC descriptor is a reliable, robust texture descriptor against noise and scale and rotation invariance.

Taking into account the cross-validation values of each method, LBP has a value of 7.69%, LTP obtains 19.23%, and RBC a cross-validation score of 11.54%. With these results, it can be said that, in terms of being independent from the training images, LBP is better than LTP and RBC. Considering the runtime to identify the pain in an input image, LBP takes around 20 ms in processing a frame, LTP around 300 ms, and RBC around 30 ms. Therefore, in terms of cross-validation score and execution time results, LBP obtains better results. However, RBC behaves much better in terms of recognition rate. In Table 1, there is a summary of the obtained results.

Table 1. A summary of texture descriptors' results.

Algorithm	Recognition Rate (%)	Cross-Validation (%)	Execution Time per Frame (ms)
LBP	89.43	**7.96**	**20**
LTP	89.43	19.23	300
RBC	**95.12**	11.54	30

Considering that typically videos work at 25–30 frames per second, it can be said that both LBP and RBC would be able to analyze all frames detected in a second, allowing the system to be integrated in a mobile app or a wearable device. However, since facial expressions do not change drastically in less than a second, the recognition process would not lose accuracy by just analyzing a few frames per second, instead of 25–30. This would also reduce workload, getting a more efficient tool in terms of speed, as a result.

Finally, in Table 2, there is a comparison between our research and some previous works. All of these works have made use of the Infant COPE database and different feature extraction methods and classifiers such as texture descriptors, deep learning methods, or supervised learning methods.

Table 2. Comparison with other works.

Article	Algorithm	Recognition Rate (%)
Brahnam et. al. [3]	PCA+SVM	82.55
Brahnam et. al. [10]	PCA+SVM	88
Mansor and Rejab [11]	LBP+Gaussian	87.74–88
	LBP+K-NN	76–80
Nanni et. al [12]	LBP+SVM	82.6
Celona and Maloni [13]	LBP+HOG+ CNN	82.95
Zamzmi et al. [14]	CNN+Strain+NB	92.71
Zamzmi et al. [15]	N-CNN	84.5
The proposed method	LBP+SVM	89.43
The proposed method	LTP+SVM	89.43
The proposed method	RBC+SVM	**95.12**

From the comparison of Table 2, it can be observed that the proposed method with Radon Barcode achieves the best recognition rate, over 10%, compared with previous works working with the same database. Therefore, it can be said that the proposed method can be used as a reliable tool to classify infant face expressions as pain or non-pain. Moreover, the time to process the algorithm makes it feasible to be implemented in a mobile app or a wearable device.

Finally, from the results in Table 2, it must be pointed out that different research that has used the same algorithms may provide different recognition rate results. This may be the result of the pre-processing stage in each work or due to the input parameters of the different feature extraction methods and/or the classifier used.

4. Conclusions

In this paper, a tool to identify infants' pain using machine learning has been implemented. The system achieves a great recognition rate when using Radon Barcodes, around 95.12%. This is the first time that RBC is used to recognize facial expressions, which proves the validity of the Radon Barcodes algorithm for the identification of emotions. In addition, as shown in Table 2, it has been proved that Radon Barcodes improved the recognition results compared to other recent proposed methods. Furthermore, the time to process frames for pain recognition with RBC makes it possible to use our system in a real mobile application.

In relation to this, we are currently working in implementing the tool in real time and designing a real wearable device to detect pain with facial images. We are beginning a collaboration with some hospitals to perform different tests and develop a prototype of the final system. Finally, we are also working with other infant databases and datasets with other ages to check the functionality and validity of the implemented tool, and the definition of a parameter to estimate the degree of pain is also under research.

Author Contributions: Conceptualization, F.A.P.; Formal analysis, H.M.; Investigation, A.M.; Methodology, F.A.P.; Resources, H.M.; Software, A.M.; Supervision, F.A.P.; Writing—original draft, A.M.; Writing—review & editing, H.M. All authors have read and agreed to the published version of the manuscript.

Funding: This work has been partially supported by the Spanish Research Agency (AEI) and the European Regional Development Fund (FEDER) under project CloudDriver4Industry TIN2017-89266-R, and by the Conselleria de Educación, Investigación, Cultura y Deporte, of the Community of Valencia, Spain, within the program of support for research under project AICO/2017/134.

Conflicts of Interest: The authors declare no conflict of interest.

References

1. Ekman, P. Facial expression and emotion. *Am. Psychol.* **1993**, *48*, 384–392. [CrossRef]
2. Pujol, F.A.; Mora, H.; Martínez, A. Emotion Recognition to Improve e-Healthcare Systems in Smart Cities. In Proceedings of the Research & Innovation Forum 2019, Rome, Italy, 24–26 April 2019; Springer: Cham, Switzerland, 2019; pp. 245–254. [CrossRef]
3. Brahnam, S.; Chuang, C.F.; Sexton, R.S.; Shih, F.Y. Machine assessment of neonatal facial expressions of acute pain. *Decis. Support Syst.* **2007**, *43*, 1242–1254. [CrossRef]
4. Roy, S.D.; Bhowmik, M.K.; Saha, P.; Ghosh, A.K. An Approach for Automatic Pain Detection through Facial Expression. *Procedia Comput. Sci.* **2016**, *84*, 99–106. [CrossRef]
5. Lucey, P.; Cohn, J.F.; Prkachin, K.M.; Solomon, P.E.; Matthews, I. Painful data: The UNBC-McMaster shoulder pain expression archive database. In *Face and Gesture 2011*; IEEE: Piscataway, NJ, USA, 2011; pp. 57–64. [CrossRef]
6. Hammal, Z.; Cohn, J.F. Automatic detection of pain intensity. In Proceedings of the 14th ACM international conference on Multimodal interaction—ICMI '12, Santa Monica, CA, USA, 22–26 October 2012; p. 47. [CrossRef]
7. Lucey, P.; Cohn, J.F.; Matthews, I.; Lucey, S.; Sridharan, S.; Howlett, J.; Prkachin, K.M. Automatically Detecting Pain in Video Through Facial Action Units. *IEEE Trans. Syst. Man Cybern. Part B Cybern.* **2011**, *41*, 664–674. [CrossRef]
8. Rodriguez, P.; Cucurull, G.; Gonzàlez, J.; Gonfaus, J.M.; Nasrollahi, K.; Moeslund, T.B.; Roca, F.X. Deep Pain: Exploiting Long Short-Term Memory Networks for Facial Expression Classification. *IEEE Trans. Cybern.* **2017**, 1–11. [CrossRef] [PubMed]
9. Ilyas, C.M.A.; Haque, M.A.; Rehm, M.; Nasrollahi, K.; Moeslund, T.B. Facial Expression Recognition for Traumatic Brain Injured Patients. In Proceedings of the 13th International Joint Conference on Computer Vision, Imaging and Computer Graphics Theory and Applications, Funchal, Madeira, Portugal, 27–29 January 2018; SCITEPRESS—Science and Technology Publications: Setúbal, Portugal, 2018; pp. 522–530. [CrossRef]
10. Brahnam, S.; Chuang, C.F.; Shih, F.Y.; Slack, M.R. Machine recognition and representation of neonatal facial displays of acute pain. *Artif. Intell. Med.* **2006**, *36*, 211–222. [CrossRef] [PubMed]

11. Naufal Mansor, M.; Rejab, M.N. A computational model of the infant pain impressions with Gaussian and Nearest Mean Classifier. In Proceedings of the 2013 IEEE International Conference on Control System, Computing and Engineering, Penang, Malaysia, 29 November–1 December 2013; pp. 249–253. [CrossRef]
12. Nanni, L.; Lumini, A.; Brahnam, S. Local binary patterns variants as texture descriptors for medical image analysis. *Artif. Intell. Med.* **2010**, *49*, 117–125. [CrossRef]
13. Celona, L.; Manoni, L. Neonatal Facial Pain Assessment Combining Hand-Crafted and Deep Features. In Proceedings of the New Trends in Image Analysis and Processing—ICIAP 2017, Catania, Italy, 11–15 September; Springer: Cham, Switzerland, 2017; pp. 197–204. [CrossRef]
14. Zamzmi, G.; Goldgof, D.; Kasturi, R.; Sun, Y. Neonatal Pain Expression Recognition Using Transfer Learning. *arXiv* **2018**, arXiv:1807.01631.
15. Zamzmi, G.; Paul, R.; Goldgof, D.; Kasturi, R.; Sun, Y. Pain assessment from facial expression: Neonatal convolutional neural network (N-CNN). In Proceedings of the 2019 International Joint Conference on Neural Networks (IJCNN), Budapest, Hungary, 14–19 July 2019; IEEE: Piscataway, NJ, USA, 2019; pp. 1–7,
16. Sun, Y.; Shan, C.; Tan, T.; Long, X.; Pourtaherian, A.; Zinger, S.; With, P.H.N.d. Video-based discomfort detection for infants. *Mach. Vis. Appl.* **2019**, *30*, 933–944. [CrossRef]
17. Lisetti, C.L.; Nasoz, F. Using Noninvasive Wearable Computers to Recognize Human Emotions from Physiological Signals. *EURASIP J. Adv. Signal Process.* **2004**, *2004*, 929414. [CrossRef]
18. Marín-Morales, J.; Higuera-Trujillo, J.L.; Greco, A.; Guixeres, J.; Llinares, C.; Scilingo, E.P.; Alcañiz, M.; Valenza, G. Affective computing in virtual reality: Emotion recognition from brain and heartbeat dynamics using wearable sensors. *Sci. Rep.* **2018**, *8*, 1–15. [CrossRef] [PubMed]
19. Miranda Calero, J.A.; Marino, R.; Lanza-Gutierrez, J.M.; Riesgo, T.; Garcia-Valderas, M.; Lopez-Ongil, C. Embedded Emotion Recognition within Cyber-Physical Systems using Physiological Signals. In Proceedings of the 2018 Conference on Design of Circuits and Integrated Systems (DCIS), Lyon, France, 14–16 November 2018; pp. 1–6. [CrossRef]
20. Chen, M.; Ma, Y.; Li, Y.; Wu, D.; Zhang, Y.; Youn, C.H. Wearable 2.0: Enabling Human-Cloud Integration in Next, Generation Healthcare Systems. *IEEE Commun. Mag.* **2017**, *55*, 54–61. [CrossRef]
21. Kwon, J.; Kim, D.H.; Park, W.; Kim, L. A wearable device for emotional recognition using facial expression and physiological response. In Proceedings of the 2016 38th Annual International Conference of the IEEE Engineering in Medicine and Biology Society (EMBC), Orlando, FL, USA, 16–20 August 2016; pp. 5765–5768, ISSN 1557-170X. [CrossRef]
22. Washington, P.; Voss, C.; Haber, N.; Tanaka, S.; Daniels, J.; Feinstein, C.; Winograd, T.; Wall, D. A Wearable Social Interaction Aid for Children with Autism. In Proceedings of the 2016 CHI Conference Extended Abstracts on Human Factors in Computing Systems, CHI EA '16, San Jose, CA, USA, 7–12 May 2016; ACM: New York, NY, USA, 2016; pp. 2348–2354. [CrossRef]
23. Kwon, J.; Kim, L. Emotion recognition using a glasses-type wearable device via multi-channel facial responses. *arXiv* **2019**, arXiv:1905.05360.
24. Dias, D.; Paulo Silva Cunha, J. Wearable Health Devices—Vital Sign Monitoring, Systems and Technologies. *Sensors* **2018**, *18*, 2414. [CrossRef]
25. Chen, W.; Dols, S.; Oetomo, S.B.; Feijs, L. Monitoring Body Temperature of Newborn Infants at Neonatal Intensive Care Units Using Wearable Sensors. In Proceedings of the Fifth International Conference on Body Area Networks, BodyNets '10, Corfu Island, Greece, 10–12 September 2010; ACM: New York, NY, USA, 2010; pp. 188–194. [CrossRef]
26. Mahmud, M.S.; Wang, H.; Fang, H. Design of a Wireless Non-Contact Wearable System for Infants Using Adaptive Filter. In Proceedings of the 10th EAI International Conference on Mobile Multimedia Communications, Chongqing, China, 13 July 2017. [CrossRef]
27. Lobo, M.A.; Hall, M.L.; Greenspan, B.; Rohloff, P.; Prosser, L.A.; Smith, B.A. Wearables for Pediatric Rehabilitation: How to Optimally Design and Use Products to Meet the Needs of Users. *Phys. Ther.* **2019**, *99*, 647–657. [CrossRef]
28. Cattani, L.; Saini, H.P.; Ferrari, G.; Pisani, F.; Raheli, R. SmartCED: An Android application for neonatal seizures detection. In Proceedings of the 2016 IEEE International Symposium on Medical Measurements and Applications (MeMeA), Benevento, Italy, 15–18 May 2016; pp. 1–6. [CrossRef]
29. Bonafide, C.P.; Jamison, D.T.; Foglia, E.E. The Emerging Market of Smartphone-Integrated Infant Physiologic Monitors. *JAMA* **2017**, *317*, 353–354. [CrossRef]

30. King, D. Marketing wearable home baby monitors: Real peace of mind? *BMJ* **2014**, *349*. [CrossRef]
31. Wang, J.; O'Kane, A.A.; Newhouse, N.; Sethu-Jones, G.R.; de Barbaro, K. Quantified Baby: Parenting and the Use of a Baby Wearable in the Wild. *Proc. Acm -Hum.-Comput. Interact.* **2017**, *1*, 1–19. [CrossRef]
32. Roofthooft, D.W.E.; Simons, S.H.P.; Anand, K.J.S.; Tibboel, D.; Dijk, M.v. Eight Years Later, Are We Still Hurting Newborn Infants? *Neonatology* **2014**, *105*, 218–226. [CrossRef]
33. Cruz, M.D.; Fernandes, A.M.; Oliveira, C.R. Epidemiology of painful procedures performed in neonates: A systematic review of observational studies. *Eur. J. Pain* **2016**, *20*, 489–498. [CrossRef]
34. Goksan, S.; Hartley, C.; Emery, F.; Cockrill, N.; Poorun, R.; Moultrie, F.; Rogers, R.; Campbell, J.; Sanders, M.; Adams, E.; et al. fMRI reveals neural activity overlap between adult and infant pain. *eLife* **2015**, *4*, e06356. [CrossRef] [PubMed]
35. Eriksson, M.; Campbell-Yeo, M. Assessment of pain in newborn infants. *Semin. Fetal Neonatal Med.* **2019**, *24*, 101003. [CrossRef] [PubMed]
36. Pettersson, M.; Olsson, E.; Ohlin, A.; Eriksson, M. Neurophysiological and behavioral measures of pain during neonatal hip examination. *Paediatr. Neonatal Pain* **2019**, *1*, 15–20. [CrossRef]
37. Stevens, B.; Johnston, C.; Petryshen, P.; Taddio, A. Premature Infant Pain Profile: Development and Initial Validation. *Clin. J. Pain* **1996**, *12*, 13. [CrossRef] [PubMed]
38. Krechel, S.W.; Bildner, J. CRIES: A new neonatal postoperative pain measurement score. Initial testing of validity and reliability. *Paediatr. Anaesth.* **1995**, *5*, 53–61. [CrossRef]
39. Lawrence, J.; Alcock, D.; McGrath, P.; Kay, J.; MacMurray, S.B.; Dulberg, C. The development of a tool to assess neonatal pain. *Neonatal Netw. NN* **1993**, *12*, 59–66. [CrossRef]
40. Grunau, R.V.E.; Craig, K.D. Pain expression in neonates: Facial action and cry. *Pain* **1987**, *28*, 395–410. [CrossRef]
41. Grunau, R.V.E.; Johnston, C.C.; Craig, K.D. Neonatal facial and cry responses to invasive and non-invasive procedures. *Pain* **1990**, *42*, 295–305. [CrossRef]
42. Peters, J.W.B.; Koot, H.M.; Grunau, R.E.; de Boer, J.; van Druenen, M.J.; Tibboel, D.; Duivenvoorden, H.J. Neonatal Facial Coding System for Assessing Postoperative Pain in Infants: Item Reduction is Valid and Feasible:. *Clin. J. Pain* **2003**, *19*, 353–363. [CrossRef]
43. Sumathi, C.P.; Santhanam, T.; Mahadevi, M. Automatic Facial Expression Analysis A Survey. *Int. J. Comput. Sci. Eng. Surv.* **2012**, *3*, 47–59. [CrossRef]
44. Kumari, J.; Rajesh, R.; Pooja, K.M. Facial Expression Recognition: A Survey. *Procedia Comput. Sci.* **2015**, *58*, 486–491. [CrossRef]
45. Arias, M.C.C.; Guinsburg, R. Differences between uni-and multidimensional scales for assessing pain in term newborn infants at the bedside. *Clinics* **2012**, *67*, 1165–1170. [CrossRef]
46. Witt, N.; Coynor, S.; Edwards, C.; Bradshaw, H. A Guide to Pain Assessment and Management in the Neonate. *Curr. Emerg. Hosp. Med. Rep.* **2016**, *4*, 1–10. [CrossRef] [PubMed]
47. Ahmed, M.; Shaukat, A.; Akram, M.U. Comparative analysis of texture descriptors for classification. In Proceedings of the 2016 IEEE International Conference on Imaging Systems and Techniques (IST), Chania, Greece, 4–6 October 2016; pp. 24–29. [CrossRef]
48. Ahonen, T.; Hadid, A.; Pietikäinen, M. Face Recognition with Local Binary Patterns. In *Computer Vision—ECCV 2004*; Pajdla, T., Matas, J., Eds.; Springer: Berlin/Heidelberg, Germany, 2004; pp. 469–481,
49. Shan, C.; Gong, S.; McOwan, P. Robust facial expression recognition using local binary patterns. In Proceedings of the IEEE International Conference on Image Processing 2005, Genoa, Italy, 11–14 September 2005; Volume 2, p. II-370, ISSN 2381-8549. [CrossRef]
50. Liu, L.; Fieguth, P.; Wang, X.; Pietikäinen, M.; Hu, D. Evaluation of LBP and Deep Texture Descriptors with a New Robustness Benchmark. In *Lecture Notes in Computer Science, Proceedings of the Computer Vision—ECCV 2016, Amsterdam, The Netherlands, 11–14 October 2016*; Leibe, B., Matas, J., Sebe, N., Welling, M., Eds.; Springer International Publishing: Cham, Switzerland, 2016; pp. 69–86. [CrossRef]
51. Tan, X.; Triggs, B. Enhanced Local Texture Feature Sets for Face Recognition Under Difficult Lighting Conditions. In *Analysis and Modeling of Faces and Gestures*; Springer: Berlin/Heidelberg, Germany, 2007; pp. 168–182. [CrossRef]
52. Hoang, T.V.; Tabbone, S. Invariant pattern recognition using the RFM descriptor. *Pattern Recognit.* **2012**, *45*, 271–284. [CrossRef]

53. Tizhoosh, H.R. Barcode annotations for medical image retrieval: A preliminary investigation. In Proceedings of the 2015 IEEE International Conference on Image Processing (ICIP), Quebec City, QC, Canada, 27–30 September 2015; pp. 818–822. [CrossRef]
54. Smyser, C.D.; Dosenbach, N.U.F.; Smyser, T.A.; Snyder, A.Z.; Rogers, C.E.; Inder, T.E.; Schlaggar, B.L.; Neil, J.J. Prediction of brain maturity in infants using machine-learning algorithms. *NeuroImage* **2016**, *136*, 1–9. [CrossRef] [PubMed]
55. Guenther, N.; Schonlau, M. Support vector machines. *Stata J.* **2016**, *16*, 917–937. [CrossRef]

applied sciences

MDPI

Article

Detection of Emotion Using Multi-Block Deep Learning in a Self-Management Interview App

Dong Hoon Shin [1], Kyungyong Chung [2] and Roy C. Park [3,*]

[1] Department of Computer Science, Kyonggi University, Suwon 16227, Korea; dhshin8222@gmail.com
[2] Division of Computer Science and Engineering, Kyonggi University, Suwon 16227, Korea; dragonhci@gmail.com
[3] Department of Information Communication Engineering, Sangji University, Wonju 26339, Korea
* Correspondence: roypark@sangji.ac.kr

Received: 6 October 2019; Accepted: 7 November 2019; Published: 11 November 2019

check for updates

Abstract: Recently, domestic universities have constructed and operated online mock interview systems for students' preparation for employment. Students can have a mock interview anywhere and at any time through the online mock interview system, and can improve any problems during the interviews via images stored in real time. For such practice, it is necessary to analyze the emotional state of the student based on the situation, and to provide coaching through accurate analysis of the interview. In this paper, we propose detection of user emotions using multi-block deep learning in a self-management interview application. Unlike the basic structure for learning about whole-face images, the multi-block deep learning method helps the user learn after sampling the core facial areas (eyes, nose, mouth, etc.), which are important factors for emotion analysis from face detection. Through the multi-block process, sampling is carried out using multiple AdaBoost learning. For optimal block image screening and verification, similarity measurement is also performed during this process. A performance evaluation of the proposed model compares the proposed system with AlexNet, which has mainly been used for facial recognition in the past. As comparison items, the recognition rate and extraction time of the specific area are compared. The extraction time of the specific area decreased by 2.61%, and the recognition rate increased by 3.75%, indicating that the proposed facial recognition method is excellent. It is expected to provide good-quality, customized interview education for job seekers by establishing a systematic interview system using the proposed deep learning method.

Keywords: self-management interview application; emotion analysis; facial recognition; image-mining; deep convolutional neural network

---

1. Introduction

Recently, Korea's youth unemployment rate has been high, to the extent that people aged 30 to 40 constitute more than half (56.7%) of the highly educated but economically inactive population (i.e., they cannot find good jobs). Accordingly, the severity of social waste through unemployment is increasing [1]. According to one analysis, many highly educated people who could have high-level careers have been produced through university education, but they are unable to find a good position right after graduation because they graduate without an appropriate interview clinic and without information and coaching on practical employment skills [2]. A situation in which they cannot work full time is problematic in job-seeking, and a common problem is that they have a lot of idle time as they go to graduate school, work as a freelancer, or work part-time due to parenting duties, despite their ability to hold down a good job. In addition, the hardest part when a job applicant seeks employment is preparing for the interviews [3]. In particular, since there are no set answers for an interview, interviewees must exhibit their capabilities differently, depending on their individual living environment and values. Also, since there are big differences among individuals (e.g., posture, eye contact, unhelpful language habits), one-on-one consulting is required, and content is needed by which students can practice their interview techniques anywhere and at any time (e.g., the night before the interview, in the train when going to an interview, and in the waiting room before the interview). The demand for online interview content is increasing, which allows a last check briefly prior to the interview [4]. Facial recognition systems can be broadly divided into face area detection and facial recognition. Face area detection determines the position of the face, size, posture, etc., in the video, and helps create a certain image for facial recognition [5,6]. Types of face detection include (1) the knowledge-based method that uses information about the typical face, (2) the feature-based method that looks for easily detected characteristics, despite changes in posture or lighting, (3) the template-matching method, which stores the basic shape of a few faces and performs a comparison with the input images, and (4) the appearance-based method, learning the face model from training images representative of the diversity in faces [7–9]. As a study for facial recognition, algorithms such as Haar, scale invariant feature transform (SIFT), ferns, modified census transform (MCT), histogram of oriented gradients (HOG), etc., are used to extract the feature factors of an image, and face analysis is actively performed based on them [10,11]. Recently, deep learning-based facial recognition has also been widely used, and a method of automatically extracting feature factors using a convolutional filter based on a convolutional neural network (CNN) has been used [12–14]. When a face is recognized using a specific factor, it is difficult to extract and select an optimal specific factor, depending on the original image state and application, and it is also difficult to determine a feature factor through various experimental and empirical factors.

We developed a self-management interview system and conducted a study on deep learning-based face analysis for emotion extraction to provide accurate interview services. Unlike the basic structure for learning the whole-face image, in this paper, a deep convolutional neural network (DCNN) method [15] for image analysis through a multi-block process helps the user learn after sampling the core facial areas, which is important for emotion analysis during face detection. The system proposed in this paper is expected to contribute to the creation of job opportunities by providing customized interview education that enables efficient interview management that is not constrained by space and time, and that provides an appropriate level of interview coaching. The figure included in this image is the author, who agreed to provide the figure.

The study is organized as follows. Section 2 describes the research related to facial recognition-based application services, and technology using facial recognition. Section 3 describes detection of user emotion using multi-block deep learning in the self-management interview application. For that purpose, also described is the image multi-block process to extract the face's feature points, plus multi-block selection and extraction of main features, and a proposed experiment with the deep learning process in face detection. Section 4 describes the proposed mobile service for real-time interview management, and Section 5 provides a conclusion.

2. Related Research

2.1. Facial Recognition-Based Application Services

Recently, various services based on facial recognition have been provided. The face recognition process is as follows. A camera captures a face image. Then, the eyes, eyebrows, nose, and mouth, which are the main factors for emotion extraction, are analyzed to extract characteristics data, and they are compared with feature data in a database provided for face analysis in facial recognition. The facial recognition technology analyzes facial expressions to determine emotional states, such as happiness, surprise, sadness, disgust, fear, confusion, etc., and is used for advertising-effect measurement, marketing, and education [16]. Founded by the Massachusetts Institute of Technology Media Lab, Affectiva released Affdex, a solution for recognizing facial expressions and identifying emotional states [17]. Figure 1 shows the Affectiva facial recognition platform. Affectiva modularizes emotion recognition-related artificial intelligence (AI) technology, distributes it through its website in the form of a software development kit (SDK), and opens it for use by various engineers and in business fields. It applies an emotion recognition solution to Tega, a robot that teaches foreign languages, and presents functions that provide appropriate content and gives rewards by understanding children's facial expressions. In addition, the facial recognition technology has been widely applied to various fields, such as locating crime suspects and lost children, and enabling mobile payments, in particular. It is used to arrest criminal suspects and locate lost children through artificial intelligence cameras attached on the chest, based on an agreement with US police [18].

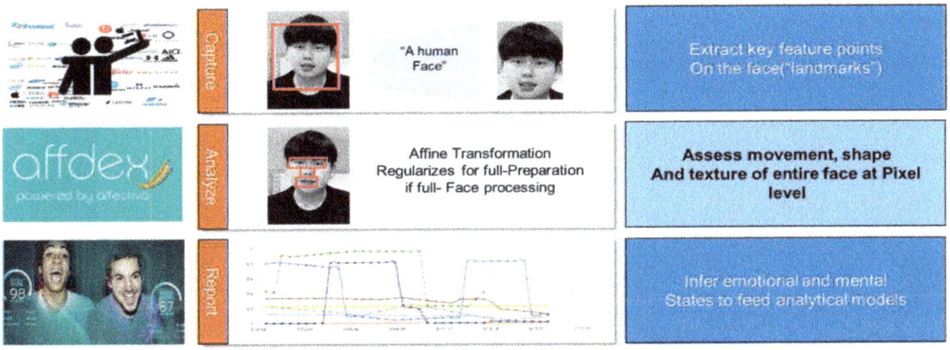

Figure 1. Facial recognition platform with respect to Affectiva [17]. * The figure included in this image is the author, who agreed to provide the figure.

2.2. *Technology Using Facial Recognition*

The AdaBoost algorithm is a technology often used for face detection, and is a method for creating strong criteria for selection by combining weak criteria, which has advantages [19,20]. This reduces the probability of drawing wrong conclusions, and increases the probability of accurately assessing problems that are difficult to judge. For facial recognition, a face area image is required. In order to increase the success rate with face detection and facial recognition, the impacts of lighting and inclination should be minimized, and images should be normalized. So, as images are normalized, the probability of errors decreases [21]. Video-based emotion recognition analyzes the characteristics of the face in a video. At first, this study used classic machine learning and computer vision. For example, the characteristics of the face were extracted based on the gradients of the face extracted from video. The characteristics were analyzed, using algorithms like a support vector machine (SVM) or random forest, to figure out the facial expression. And yet, there is a singularity effect according to the surrounding background or the illumination intensity of the video. In addition, accuracy is greatly affected by the angle of the face. Figure 2 shows a facial recognition algorithm using a face database. The dataset used in the early stages was secured in a limited environment; however, videos that contain everyday situations are in the dataset [22].

Figure 3 shows how to recognize a face in a three-dimensional (3D) image [23,24]. The flow of the method can be separated from the training phase and the test phase. In the training phase, face data are collected from 3D images, and pretreatment is performed to obtain a clean 3D face without bending. Preprocessed data include facial features from a feature extraction system [25]. The features, such as extracted face data, are stored in a feature database [26,27]. Next, in the test phase, the entered target faces are the same as those used in the training phase and during the 3D face data collection, pretreatment, and feature-extraction steps. In the feature-matching phase, match scores are calculated by comparing the target face with the database saved in the training phase. When the match score is judged to be high enough, this algorithm determines that the target face has been recognized. Also, facial recognition technology is used in the augmented reality field. It is a method to extract feature points, and draws the coordinate plane of the face to calculate the position, to make a 3D effect of the product, and overlap it onto the face [28,29].

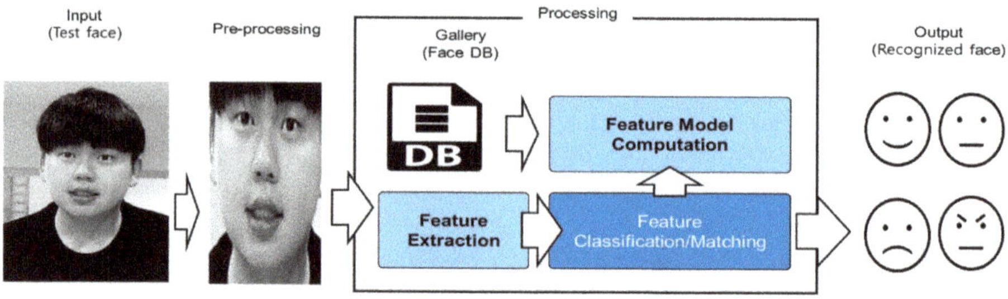

Figure 2. Facial recognition algorithm using a face database. * The figure included in this image is the author, who agreed to provide the figure.

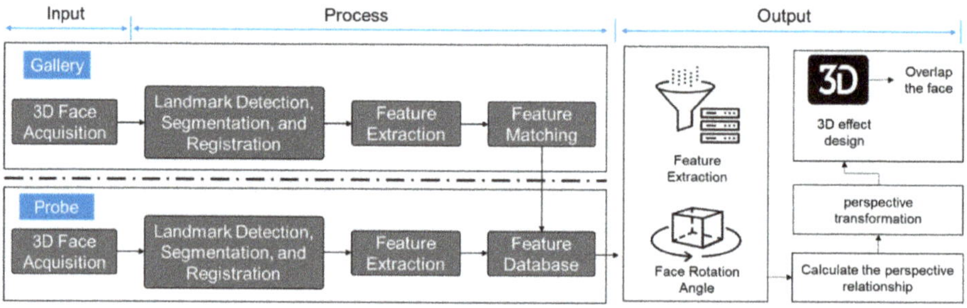

Figure 3. A 3D facial recognition of augmented reality system.

3. Detection of Emotions Using Multi-Block Deep Learning in Self-Management Interviews

Figure 4 shows the whole process of the system described in this paper. First, we proceed with facial recognition, where features are extracted. Multi-block sampling is performed by extracting feature points from the recognized faces. Sampled data are extracted through deep learning based on a DCNN. Analysis is conducted based on the extracted emotions, and the analyzed data are managed by the interview system proposed in this paper. Interview management is done through the application itself. The CAS-PEAL face database is used for facial recognition, and the Cohn-Kanade database is used for emotion extraction.

3.1. *Image Multi-Block Process for Face Main Point Extraction*

In this paper, we developed a self-management interview system and conducted a study on deep learning-based face analysis for emotion extraction to provide accurate interview services. Unlike the basic structure for learning the whole-face image, the deep learning method proposed in this paper is a model that helps the user learn from images of multi-block core areas, such as the eyes, nose, and mouth, which are important factors for emotion analysis during face detection. The proposed learning structure of the DCNN consists of a multi-block process of entered face images and multi-block deep learning. In the multi-block process, the input image is blocked based on multiple AdaBoost. The multi-block deep learning model is executed by considering the sizes of the original image and of the sampled image that is blocked for area extraction. When both processes are completed, the whole face image and the sampled multi-block image have been learned, making it possible to use them during the emotion detection stage afterwards. The recognition process of the multi-block deep learning algorithm consists of a multi-block process, multi-block selection, and a multi-block deep performance process. In the existing deep learning model, facial recognition utilizes the whole face, which causes a problem in that areas such as the eyes, nose, and mouth (the key factors for analyzing emotions) are not recognized correctly. In this paper, therefore, the recognition rate was improved by extracting the specific parts of a face image required for emotion extraction by the multi-block method. In particular, if the multi-block is large or small in the blocking process, features of the main areas cannot be extracted accurately, which causes large errors in recognition and learning.

In this paper, multiple AdaBoost was used to carry out sampling by setting the optimal blocking. Figure 5 shows the process of detection and classification with multiple AdaBoost. Multiple AdaBoost creates a stronger classifier by combining weak classifiers, allows a weak classifier to determine whether the image is a face or not when there is a certain purpose. It is designed to select a feature of a rectangular shape with the fewest errors in order to let a weak classifier use the fewest improperly classified training videos, and, in turn, have the optimum threshold classification function.

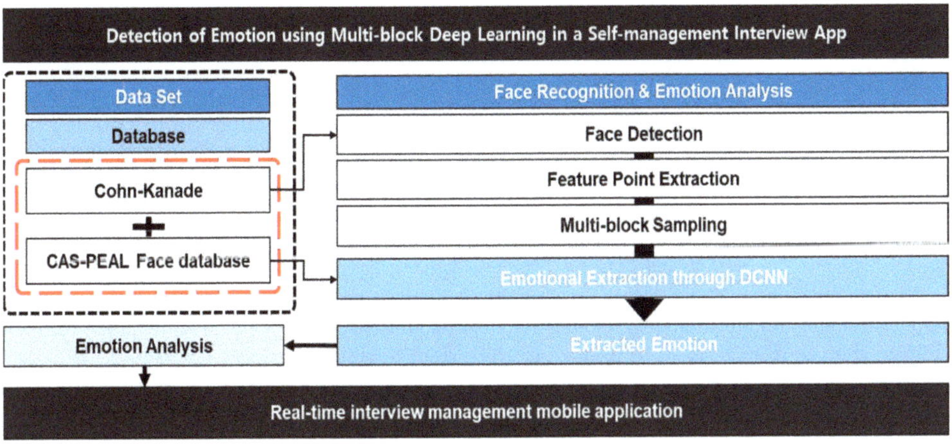

Figure 4. System-wide process for interview management.

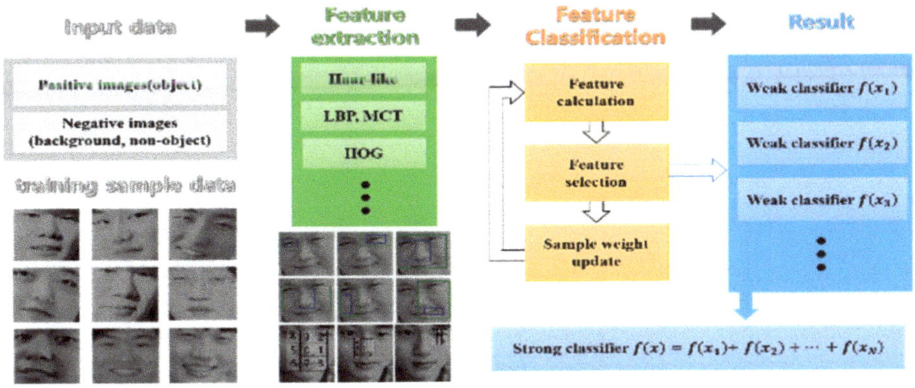

Figure 5. The process of detection and classification with multiple AdaBoost. * The figure included in this image is the author, who agreed to provide the figure.

For this process, training images and sample images were required, so by using the CAS-PEAL face database, our database included 99,594 images with a variety of poses, expressions, and lighting levels from 1040 individuals (595 male and 445 female). Domains of faces to be extracted were defined as positive (object) samples, while images other than a face were defined as negative (non-object, background) samples. Also, we use the Cohn-Kanade database to analyze perceived facial emotions from data in this database that include 486 sequences from 97 poses [30,31]. At this time, positive images must have pixels of the same size, and detection should be made by aligning the positions of eyes, noses, and mouths so they are the same as much as possible. Learning data should include information on whether the image belongs in the positive or negative category. In addition, features for distinguishing a face from the background are also required. Such features could be presented as a classifier to distinguish/classify an object. Since these features are a base classifier and a candidate for a weak classifier, it was necessary to decide how many times the process of weak classifier selection should be repeated [32,33]. In other words, it was necessary to determine how many weak classifiers should be combined into one stronger classifier, and to select one feature having the best performance in classifying training samples by class and to calculate a weak classifier for the corresponding iteration [34]. Therefore, we used a weighted linear combination of T weak classifiers, as shown in Equation (1).

$$\mathrm{E}(x) = a_1 e_1(x) + \ldots + a_T e_T(x) = \sum_{t=1}^{T} a_t e_t(x) \tag{1}$$

E: final strong classifier,
e: weak classifier,
a: weighted of weak classifier,
t: iteration round (1,2, … ,T).

3.2. Multi-Block Selection and Extraction of Main Area Features

The multi-block selection process selects blocks to be used for actual recognition among the multi-blocks previously delivered through the feature numerical analysis. For accurate emotion analysis, the user's eyes, nose, and mouth, which are the main feature points, should be clearly identified, and they can be classified according to the degree of rotation of the face. If there is no information on specific points in the whole image, the rotation information should be detected during the multi-block process. Figure 6 shows the whole facial recognition and emotion analysis process.

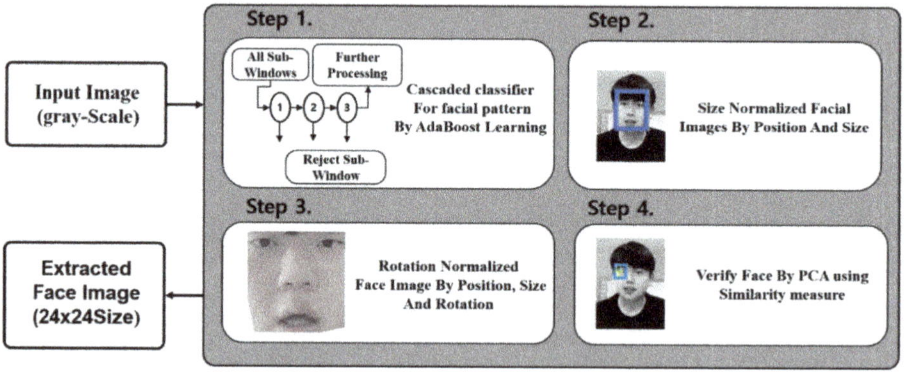

Figure 6. The whole facial recognition and emotion analysis process. * The figure included in this image is the author, who agreed to provide the figure.

Face detection was made by moving a 24 × 24 pixel block; for simple patterns in multiple AdaBoost learning, basic patterns were used. In addition, the number of simple detectors to be searched by the learning process was selected as 160, and the learned detectors became serialized, in turn enhancing the processing speed. The learned detectors were serialized into 10 stages in which 16 learned detectors belong in an arbitrary manner. Parameters for each stage were adjusted, and as for images in multiple AdaBoost learning, 24 × 24 resolution was used. For detection by size, the input images were classified (based on the degree of down-sampling) into three types, and the face was detected from among the down-sampled images. In detection by rotation, facial images rotated at −5° to +5°, +15° to +25°, and −15° to −25° were learned by AdaBoost. Then, by using the serialized detector, they were each analyzed and, in order of detection, rotation of the face was classified. The detected faces were classified into nine types, and the information about the locations of the detected faces was provided as well. Figure 7 shows the face image–detection process. For face detection, 80 × 60 down-sampled images were used for detecting a large face, 108 × 81 down-sampled images were used for a medium-sized face, and 144 × 108 down-sampled images for a small face. The sequence of detection by size was selected to enhance the detection speed and was done as follows: Detection of 80 × 60 down-sampled images was first, followed by the 108 × 81 down-sampled images, and then, the 144 × 108 images. If a detected face overlapped the block detected in the face from the down-sampled image in the preceding step, that detection was not valid. The input image was searched for among the down-sampled images, and when it was detected, a block of the face from the detected image was cut and then normalized to the predesigned size and passed to the next process. At the time, principal component analysis (PCA) was used to measure similarity with the input image, verifying the face. This is a process of rotating an image by using the verified information on rotation of the face, until the rotation of the face in the image becomes almost zero. When the rotation of the face was verified to be between +15° and +25°, the face was rotated by −20°, and when the rotation of the face was between −15° and −25°, the face was rotated by +20°.

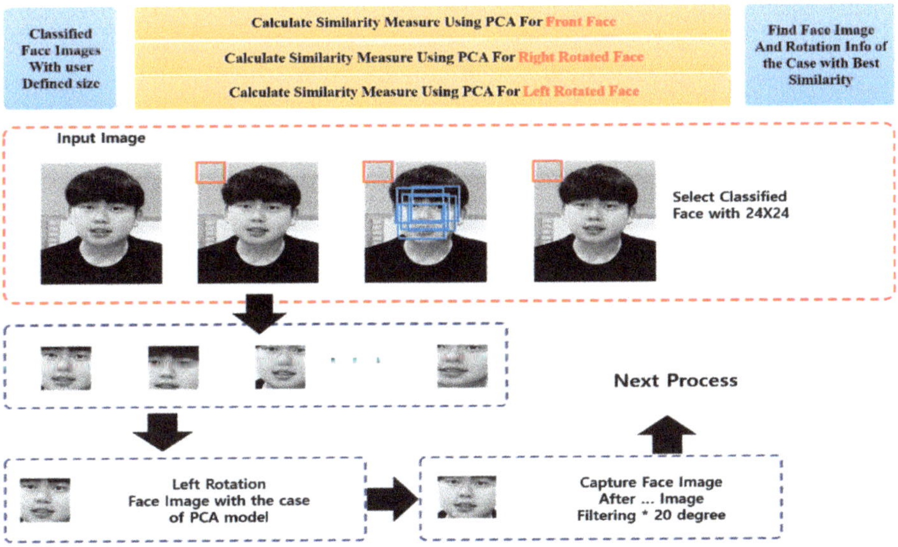

Figure 7. The face image detection process. * The figure included in this image is the author, who agreed to provide the figure.

When the user's face was extracted in the aforementioned process, the positions of the eyes and nose should be extracted. The patterns for the person's eyes could be extracted by using the facial image obtained through the face detection process. Eyes and nose extraction can be classified into three stages. The first stage was to designate a region to search for the eyes in the facial image obtained by the face detector. From this stage, it was possible to roughly estimate the position of the eyes, even if they were not precise, and such an estimated position could be defined as a certain domain. In the second stage, the region for the eyes must be clearly defined, as shown in Figure 8 After defining the eyes region, we used multiple AdaBoost to determine 12 × 12 pixel eye images and 12 × 12 pixel non-eye images to prepare the serialized eye detector. This was to classify these eyes from other eyes.

Then, AdaBoost went through a process of detecting block images that had the eyes in the designated region. The last stage was to use PCA, trained with eye images, to measure the similarity of each eye image and to select the image with the highest similarity. As shown in Figure 8, the position of the eyes could be defined as the center point of a verified eyes image.

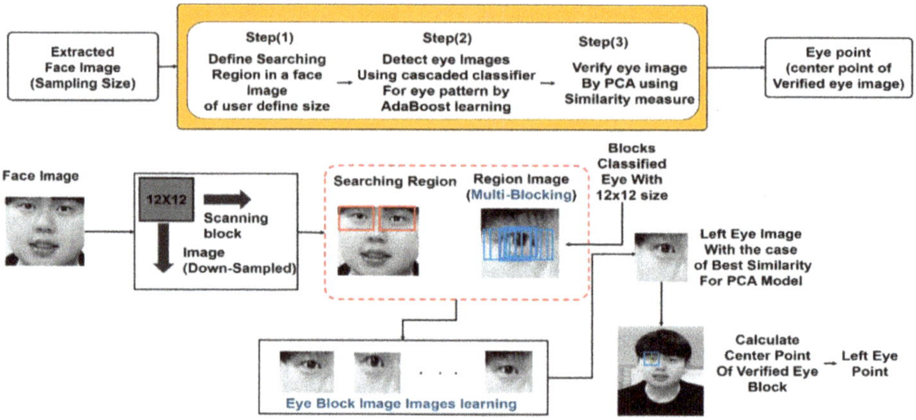

Figure 8. The eye detection process. * The figure included in this image is the author, who agreed to provide the figure.

In order to detect a nose's location, it was necessary to designate a nose search region on the face image acquired during the face search, which is the same process as required for the eye search. Although the exact location of the nose cannot be specified, a rough location can be estimated, and the predictive value of the location of the nose can be defined for certain regions. Figure 9 shows the nose detection process.

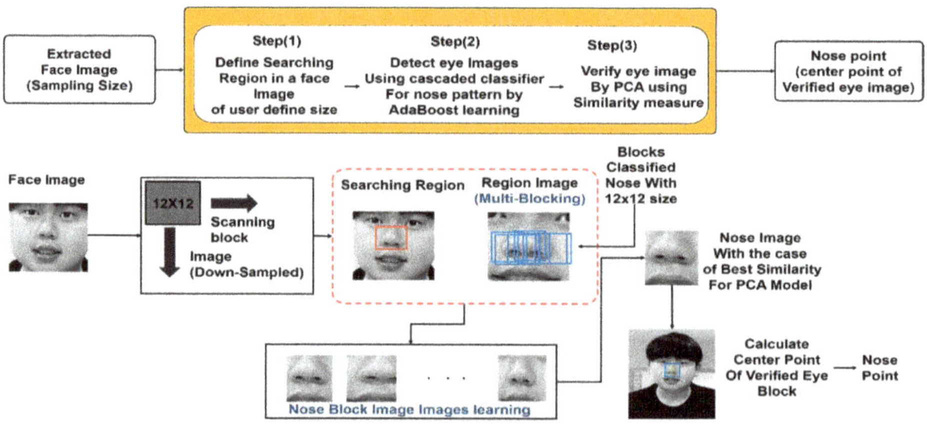

Figure 9. The nose detection process. * The figure included in this image is the author, who agreed to provide the figure.

In order to determine whether the image was actually a nose or not, multiple AdaBoost was used to learn 12 × 12 nose images and non-nose images in order to create a serialized nose detector and go through the process of finding the block images that were detected as noses within the defined region in the first step. The last step was to calculate the similarity of the nose images acquired during the second step, and compare nose images to find one with the best similarity. As with the eye image search process, the nose location was the center point of a verified nose image. After the detection of eyes and nose locations, the face normalization process was followed. Face normalization is a process to calculate the accurate location, size, and rotation of the face using the locations of both the eyes and the nose. The face image was warped to ensure consistency among the different forms of a face. The actual emotion images can be created by finding the eyes and the nose in the image and by going through image warping based on that information. At this time, the size of the normalized images and the location of each part may vary depending on the design of the recognizer.

3.3. Face Detection Using the Multi-Block Deep Learning Process

For a deep learning model of the proposed user emotion extraction, this experiment extracted emotions using a DCNN based on the multi-blocked sample images of the major face areas, and the images with completed feature extractions, which was intended to minimize the performance time from entry, and the classification of the images. Figure 10 shows the multi-block deep learning structure proposed in this study. The emotion model was extracted by delivering a block target that included information about the features from the images learned by the DCNN in the multi-block and block selection stages. A convolution operation was conducted between the original images and the multi-block images extracted by sampling. This brought into relief the features of the major face areas for the extraction of emotions through the feature extraction filter. The filter coefficient for the feature extraction filter was set to a random value in the early stages, and was then set to the optimal filter coefficient with the least error rate through learning. Next, the process of reducing the images was executed, analyzing the features of the extracted images, and filtering the optimum features. At this time, the general DCNN launched a method for minimizing the cross-entropy loss function so it can be similar to the softmax result from image data entered from the multi-block and feature extraction stages.

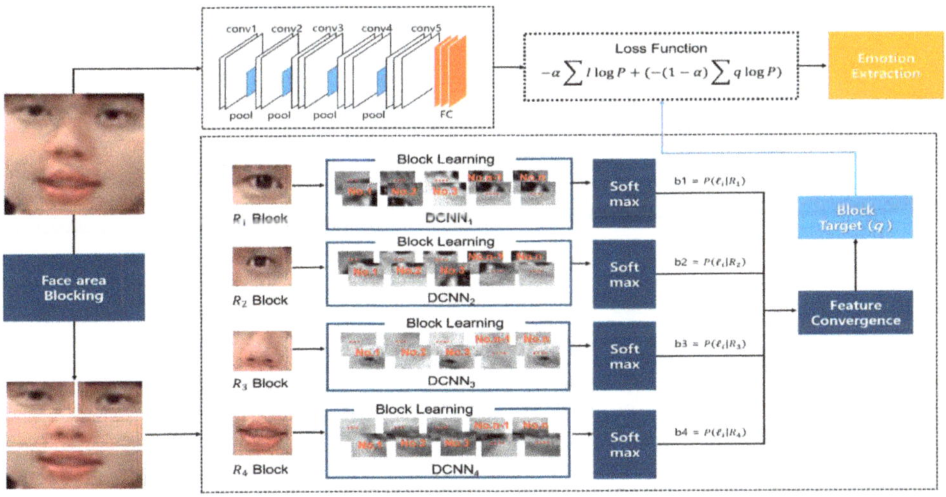

Figure 10. The proposed multi-block deep learning structure. * The figure included in this image is the author, who agreed to provide the figure.

This study defines two cross-entropy loss functions like those in Equations (2) and (3) to deliver knowledge: In Equation (2), loss function L_1 is a cross-entropy function based on a recognition result error for the label. In Equation (3), loss function L_2 is the cross-entropy function representing the error with the block target representing the predicted probability value of the DCNN.

$$L_1 = -\sum_{n=1}^{|V|} H(y = n) \times logP(y = n|x;\ \theta) \quad (2)$$

$$L_2 = -\sum_{n=1}^{|V|} q(y = n|x;\ \theta_E) \times logP(y = n|x;\ \theta) \quad (3)$$

In the formula, q is the softmax probability value formed by learning the features of multi-blocked images, while $P(y = n|x; \theta)$ is the probability the DCNN learned by utilizing the features of the whole images, n is the index of the feature category, and $|v|$ is the total number of classes. This study used both the knowledge block target containing the feature information of the multi-block images delivered by the DCNN while learning, and the existing true class target value so as to allow learning everything. It extracted accurate emotions, utilizing feature extraction of the whole image area and the features of the blocked images of the key areas, giving a different weighted value to each of the two loss functions, L_1 and L_2, as seen in Equation (4):

$$L = \alpha \times L_1 + (1 - \alpha) \times L_2,\ 0 < \alpha < 1 \quad (4)$$

In addition, for face area detection and estimation analysis, the CAS-PEAL face database was employed. The learning data in the database used consisted of classes of facial expression information for a total of 1040 persons, and consisted of a total of 1240 sheets of images for each class. The data for deep learning was composed of 10 sheets per emotion class, with noise added to the learning data. On the other hand, the AlexNet [35] structure, which is used a lot for facial recognition, was selected for comparison with the deep learning method proposed in this paper. Figure 11 shows a comparison of emotion recognition accuracy with the proposed method against the accuracy with AlexNet. The facial expressions were recognized through extraction of the entire face area and the main areas, and the accuracy of extracting emotions according to the expressions was compared, with the proposed method showing accuracy about 3.75% higher.

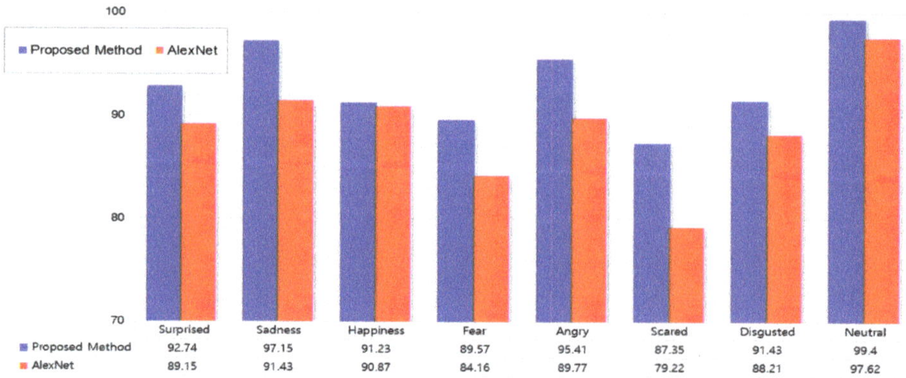

Figure 11. Comparison of emotion recognition accuracy of the proposed model and AlexNet.

Figure 12 shows the distribution of the times required to extract the face area, and the results of face area detection. As a result of one experiment, the proposed method had a faster processing time and a lower error rate than the basic method that did not go through smoothing. In addition, as the dispersion of the processing time was only a little, it turned out to be a normalization method suitable for real-time processing.

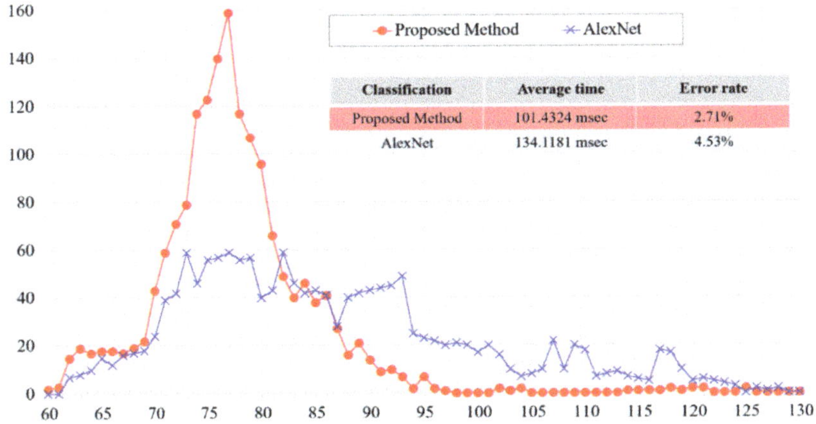

Figure 12. Face area detection time (in milliseconds).

Figure 13 shows the distribution of the processing time to detect the eye area, and the results of eye area detection. In eye area detection, the distribution of the processing time was not affected by normalization; however, there was a difference in the error rate. As a result of the experiment, the proposed method was deemed excellent.

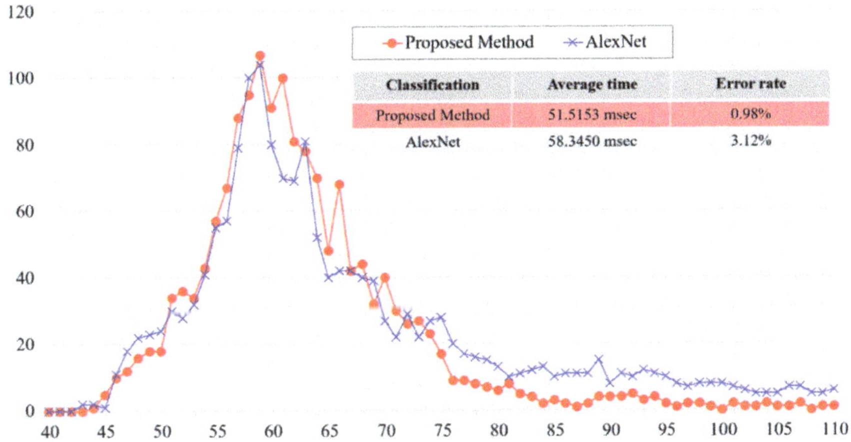

Figure 13. Eye area detection time (in milliseconds).

Figure 14 shows the distribution of the processing time to detect the nose area, and the results of nose area detection. As with eye area detection, the proposed method showed excellent performance in terms of average processing time and error rate.

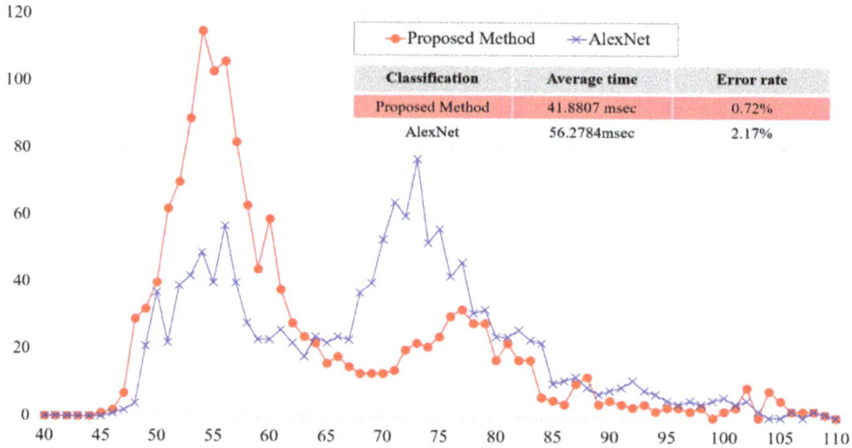

Figure 14. Nose area detection time (in millisecconds).

4. Mobile Service for Real-Time Interview Management

The self-management interview system was developed as a mobile application for smooth interview coaching. When the interview app is used, a real-time video is taken and transmitted to the server. At this time, the person's emotional state is presented through voice and facial recognition in the video, and real-time coaching is provided accordingly. In addition, including various types of interview coaching content and self-diagnosis programs, it is an effective system for speech practice as well as interviews. Figure 15 shows the image-analysis algorithm-based emotion matching. As for the image-analysis algorithm, the faces and eyes were detected, using an Extensible Markup Language (XML) classifier, and based on the detected images, emotions were extracted from a comparative analysis by the CAS-PEAL face database and Sort image. The figure included in this image is the author, who agreed to provide the figure.

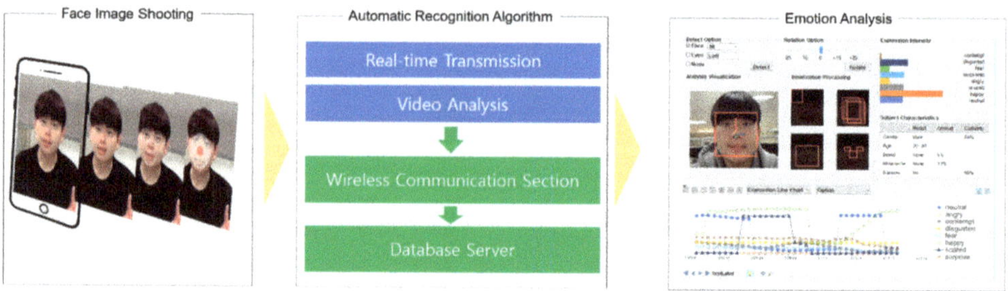

Figure 15. The structure of the interview management system. * The figure included in this image is the author, who agreed to provide the figure.

Figure 16 shows a system that analyzes images by capturing one frame after dividing a video into frame units. System functions include video playback, analysis visualization, recognition options, rotation options, binary processing, curve graph representation of emotions, object feature analysis, etc. After capturing the video, the user selects the part to be recognized with the recognition option and then recognizes that part through a binarization process. The binarization function finds the feature points of the image. There may be a rotated face in the captured image, so there is also an option that rotates the face to the correct position. This function offers a selection range of −25 to +25 degrees. There are eight emotions for analyzing a person's feelings through the recognition function: Neutral (usual expression), contempt, disgust, anger, happiness, surprise, sadness, and fear. There is also a function that graphs the emotions in each image captured from the video. Analysis of the image in Figure 16 confirms the person is happy. Object characterization analysis shows the gender and age, and features like a mustache, beard, and eyeglasses. According to the analysis, the image in the current frame is male, 20–30 years old, without a mustache or beard, and no glasses. A screen shot from the facial recognition and emotion analysis results of the interview management system is shown in Figure 17.

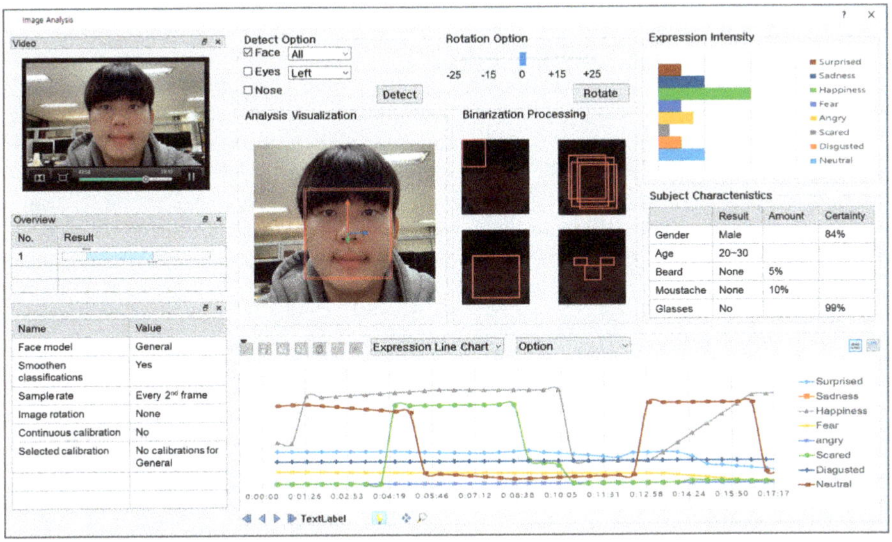

Figure 16. The real-time interview management system. * The figure included in this image is the author, who agreed to provide the figure.

Figure 17. Facial recognition and emotion analysis results from the interview management system. * The figure included in this image is the author, who agreed to provide the figure.

For the mobile service configured in this study, an application was developed utilizing Android Studio 9 (Pie) on an Intel Core i7-4770 CPU at 3.40 GHz, with 16 GB of RAM running the Windows 10 Enterprise 64-bit environment. The figure included in this image is the author, who agreed to provide the figure. For the real-time interview and automatic coaching service, an app was configured that has a server for interview management, a module for automatic coaching based on the interview when a user uses the service, and a user interface for the relevant services. When the user touches each button in the real-time interview management mobile application, including voice evaluation, interview evaluation, and comprehensive interview from the main screen, that input is passed to the service use information page, providing values for pronunciation, interview, and coaching, for the function interviewCode. On the service use information page, the value of interviewCode is forwarded as an intent that is distinguished as a value for each variable and is displayed, applying a message image for the corresponding voice evaluation, interview evaluation, comprehensive interview, and start button.

Splash screens for the facial recognition and emotion analysis results of self-managed interviews are shown in Figure 18. Once the interview evaluation begins, for evaluation questions, the application calls up the interview question API(Application Programming Interface) in the server, brings up the index of the relevant questions, the content of the questions, and information about the company that set the questions, and displays them in the application view. This was designed so that, once recording begins, the application calls the Android internal camera and conducts image recording and voice recording for encoding, so that both image and voice are included in the video. When the recording ends, the file-upload API in the server is immediately run to upload the user information, question index, and video file to the server, and once uploading is completed, the analysis procedure is launched through a module. On the module analysis information page, at regular intervals, the application continuously calls up the module analysis results API in the server. When the module analysis is completed, the user moves to the interview evaluation results page. Then, with the values coming from the module, the result is displayed in percentages of the emotions (including neutral, contempt, disgusted, angry, happy, surprised, scared, and fear) in the criteria for analysis.

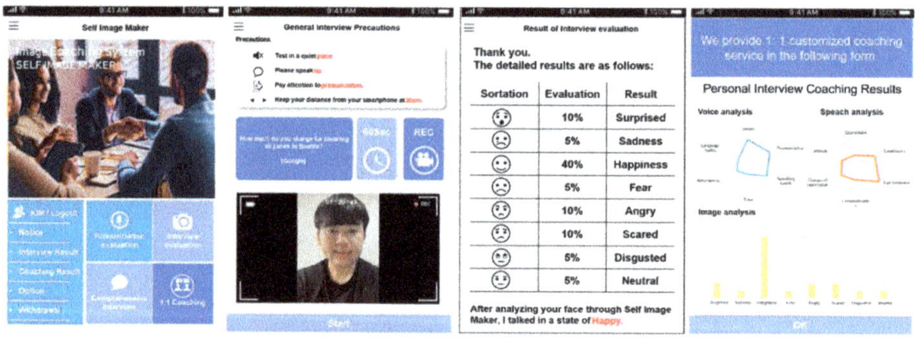

Figure 18. The real-time interview management mobile application. * The figure included in this image is the author, who agreed to provide the figure.

5. Conclusions

In this paper, we developed a self-management interview system and conducted a study on deep learning-based face analysis for emotion extraction to provide an accurate interview evaluation service. A self-management interview system was developed as a mobile application for smooth interview coaching. When the interview service is used, a real-time video is recorded and transmitted to the server. At this time, the person's emotional state is presented through voice and facial recognition from the video, and real-time coaching is provided accordingly. In addition, including a variety of interview coaching content and self-diagnosis programs, the proposed system is effective for speech practice as well as interview practice. Unlike the basic structure for recognizing a whole-face image, the deep learning method for image analysis in this system helps the user learn after sampling the core areas that are important for sentiment analysis during face detection through a multi-block process. In the multi-block process, multiple AdaBoost is used to perform sampling. After sampling, an XML classifier is used to detect the main features, which are set at threshold values to remove elements that interfere with facial recognition. In addition, the extracted images are detected by using the CAS-PEAL face database to classify eight emotions (e.g., neutral, contempt, disgusted, angry, happy, surprised, scared, and fear), and services are provided through the application. In the experiment results, facial expressions were recognized through extraction of the entire face area and the main areas. The accuracy from extracting emotions based on the recorded expressions was compared, and the extraction time of the specific areas was decreased by 2.61%, and the recognition rate was increased by 3.75%, indicating that the proposed facial recognition method is excellent. The extracted emotions are provided through an interview management app, and users can efficiently access the interview management system based on them. We believe the interview coaching application will be utilized to provide an interview education that matches students with employment coaches, and it will provide quality job interview–education content for students in the future. The system proposed in this paper is expected to contribute to the creation of job opportunities by providing customized interview education that enables efficient interview management, is not constrained by space and time, and provides an appropriate level of interview coaching.

Author Contributions: K.C. and R.C.P. conceived and designed the framework. D.H.S. implemented Multi-Block Deep Learning for a Self-Management Interview App. R.C.P. and D.H.S. performed experiments and analyzed the results. All authors have contributed in writing and proofreading the paper.

Funding: This research was funded by a National Research Foundation of Korea (NRF) grant funded by the Korea government (2019R1F1A1060328).

Acknowledgments: We appreciate very much the author and researchers who agreed to provide the images used in this paper.

Conflicts of Interest: The authors declare no conflict of interest.

References

1. Jang, H. The effectiveness of labor market policies on youth employment. *Korean Public Adm. Rev.* **2017**, *51*, 325–358. [CrossRef]
2. Na, E.M. The Identity of the Employment-oriented Genre and the Design of Self-introduction writing and interview. *Korean J. Lit. Res.* **2018**, *9*, 131–159.
3. Park, J.H.; Lee, K.J.; Lee, S.W. The Effects of Local Industrial Structures on the Probability of Youth Employment. *J. Korean Reg. Dev. Assoc.* **2018**, *30*, 133–159.
4. Choi, C.S. Effects of Position Interdependency and Competitiveness on Communication Strategy Selection in Groupdiscussion Employment Interviews. *J. Commun. Res.* **2019**, *56*, 330–371.
5. Kan, H.; Xiang, W.; Zhenan, S.; Tieniu, T. Wasserstein CNN: Learning Invariant Features for NIR-VIS Face Recognition. *IEEE Trans. Pattern Anal. Mach. Intell.* **2019**, *41*, 1761–1773.
6. Zhou, L.; Li, W.; Du, Y.; Lei, B.; Liang, S. Adaptive illumination-invariant face recognition via local nonlinear multi-layer contrast feature. *J. Vis. Commun. Image Represent.* **2019**, *64*, 102641. [CrossRef]
7. Cai, Y.; Lei, Y.; Yang, M.; You, Z.; Shan, S. A fast and robust 3D face recognition approach based on deeply learned face representation. *Neurocomputing* **2019**, *363*, 375–397. [CrossRef]
8. Lia, C.; Huang, Y.; Yu, X. Dependence Structure of Gabor Wavelets based on Copula for Face Recognition. *Expert Syst. Appl.* **2019**, *137*, 453–470. [CrossRef]
9. Shakeel, M.S.; Lam, K.-M.; Lai, S.-C. Learning sparse discriminant low-rank features for low-resolution face recognition. *J. Vis. Commun. Image Represent.* **2019**, *63*, 102590. [CrossRef]
10. Wu, Y.; Ji, Q. Facial Landmark Detection: A Literature Survey. *Int. J. Comput. Vis.* **2019**, *127*, 115–142. [CrossRef]
11. Corneanu, C.A.; Simon, M.O.; Cohn, J.F.; Guerrero, S.E.; Oliu, M.; Escalera, S. Survey on RGB, 3D, Thermal, and Multimodal Approaches for Facial Expression Recognition: History, Trends, and Affect-Related Applications. *IEEE Trans. Pattern Anal. Mach. Intell.* **2016**, *38*, 1548–1568. [CrossRef] [PubMed]
12. Iqbal, M.; Sameem, M.S.; Naqvi, N.; Kanwal, S.; Ye, Z. A Deep Learning Approach for Face Recognition based on Angularly Discriminative Features. *Pattern Recognit. Lett.* **2019**, *128*, 414–419. [CrossRef]
13. Guo, G.; Zhang, N. A Survey on Deep Learning based Face Recognition. *Comput. Vis. Image Underst.* **2019**, *189*, 102805. [CrossRef]

14. Elmahmudi, A.; Hassan, U. Deep Face Recognition using Imperfect Facial Data. *Future Gener. Comput. Syst.* **2019**, *99*, 213–225. [CrossRef]
15. Mayya, V.; Pai, R.M.; Pai, M.M. Automatic Facial Expression Recognition Using DCNN. *Procedia Comput. Sci.* **2016**, *93*, 453–461. [CrossRef]
16. Chung, K.; Park, R.C. Cloud based U-healthcare Network with QoS Guarantee for Mobile Health Service. *Clust. Comput.* **2019**, *22*, 2001–2015. [CrossRef]
17. Affectiva. Available online: https://www.affectiva.com/ (accessed on 9 October 2019).
18. Chung, K.; Park, R.C. Chatbot-based heathcare service with a knowledge base for cloud computing. *Clust. Comput.* **2019**, *22*, 1925–1937. [CrossRef]
19. Xu, Y.; Li, Z.; Yang, J.; Zhang, D. A Survey of Dictionary Learning Algorithms for Face Recognition. *IEEE Access* **2017**, *5*, 8502–8514. [CrossRef]
20. Fu, S.; He, H.; Hou, Z.G. Learning Race from Face: A Survey. *IEEE Trans. Pattern Anal. Mach. Intell.* **2014**, *36*, 2483–2509. [CrossRef] [PubMed]
21. Chung, K.; Park, R.C. PHR Open Platform based Smart Health Service using Distributed Object Group Framework. *Clust. Comput.* **2016**, *19*, 505–517. [CrossRef]
22. Gong, D.; Li, Z.; Huang, W.; Li, X.; Tao, D. Heterogeneous Face Recognition: A Common Encoding Feature Discriminant Approach. *IEEE Trans. Image Process.* **2017**, *26*, 2079–2089. [CrossRef] [PubMed]
23. Violante, M.G.; Marcolin, F.; Vezzetti, E.; Ulrich, L.; Billia, G.; Di Grazia, L. 3D Facial Expression Recognition for Defining Users' Inner Requirements-An Emotional Design Case Study. *Appl. Sci.* **2019**, *9*, 2218. [CrossRef]
24. Wang, J.; Yin, L.; Wei, X.; Sun, Y. 3D facial expression recognition based on primitive surface feature distribution. In Proceedings of the 2006 IEEE Computer Society Conference on Computer Vision and Pattern Recognition, New York, NY, USA, 17–22 June 2006; Volume 2, pp. 1399–1406.
25. Zhou, S.; Xiao, S. 3D face recognition: A survey. *Hum. Cent. Comput. Inf. Sci.* **2018**, *8*, 35. [CrossRef]
26. Yin, L.; Wei, X.; Sun, Y.; Wang, J.; Rosato, M.J. A 3D facial expression database for facial behavior research. In Proceedings of the 7th International Conference on Automatic Face and Gesture Recognition, Southampton, UK, 10–12 April 2006; pp. 211–216.
27. Kim, J.C.; Chung, K. Prediction model of user physical activity using data characteristics-based long short-term memory recurrent neural networks. *Ksii Trans. Internet Inf. Syst.* **2019**, *13*, 2060–2077.
28. Kim, J.C.; Chung, K. Associative feature information extraction using text mining from health big data. *Wirel. Pers. Commun.* **2019**, *105*, 691–707. [CrossRef]
29. Yoo, H.; Chung, K. Mining-based lifecare recommendation using peer-to-peer dataset and adaptive decision feedback. *Peer-to-Peer Netw. Appl.* **2018**, *11*, 1309–1320. [CrossRef]
30. Gao, W.; Cao, B.; Shan, S.; Chen, X.; Zhou, D.; Zhang, X.; Zhao, D. The CAS-PEAL Large-Scale Chinese Face Database and Baseline Evaluations. *IEEE Trans. Syst. Man Cybern.* **2008**, *38*, 149–161.
31. Chung, K.; Yoo, H.; Choe, D.; Jung, H. Blockchain network based topic mining process for cognitive manufacturing. *Wirel. Pers. Commun.* **2019**, *105*, 583–597. [CrossRef]
32. Yoo, H.; Chung, K. Heart rate variability based stress index service model using bio-sensor. *Clust. Comput.* **2018**, *21*, 1139–1149. [CrossRef]
33. Song, C.W.; Jung, H.; Chung, K. Development of a medical big-data mining process using topic modeling. *Clust. Comput.* **2019**, *22*, 1949–1958. [CrossRef]
34. Baek, J.W.; Kim, J.C.; Chun, J.; Chung, K. Hybrid clustering based health decision-making for improving dietary habits. *Technol. Health Care* **2019**, *27*, 459–472. [CrossRef] [PubMed]
35. Lin, L.; Yingzi, T. Detection of Cabinet in Equipment Floor based on AlexNet and SSD Model. *J. Eng.* **2019**, *2019*, 605–608.

applied sciences

Article

Cost-Effective CNNs for Real-Time Micro-Expression Recognition

Reda Belaiche *, Yu Liu, Cyrille Migniot, Dominique Ginhac and Fan Yang

ImViA EA 7535, University Bourgogne Franche-Comté, 21000 Dijon, France; Yu_liu@etu.u-Bourgogne.fr (Y.L.); Cyrille.Migniot@u-bourgogne.fr (C.M.); Dominique.Ginhac@u-bourgogne.fr (D.G.); fanyang@u-bourgogne.fr (F.Y.)

* Correspondence: Reda.Belaiche@u-bourgogne.fr

Received: 5 June 2020; Accepted: 16 July 2020; Published: 19 July 2020

Abstract: Micro-Expression (ME) recognition is a hot topic in computer vision as it presents a gateway to capture and understand daily human emotions. It is nonetheless a challenging problem due to ME typically being transient (lasting less than 200 ms) and subtle. Recent advances in machine learning enable new and effective methods to be adopted for solving diverse computer vision tasks. In particular, the use of deep learning techniques on large datasets outperforms classical approaches based on classical machine learning which rely on hand-crafted features. Even though available datasets for spontaneous ME are scarce and much smaller, using off-the-shelf Convolutional Neural Networks (CNNs) still demonstrates satisfactory classification results. However, these networks are intense in terms of memory consumption and computational resources. This poses great challenges when deploying CNN-based solutions in many applications, such as driver monitoring and comprehension recognition in virtual classrooms, which demand fast and accurate recognition. As these networks were initially designed for tasks of different domains, they are over-parameterized and need to be optimized for ME recognition. In this paper, we propose a new network based on the well-known ResNet18 which we optimized for ME classification in two ways. Firstly, we reduced the depth of the network by removing residual layers. Secondly, we introduced a more compact representation of optical flow used as input to the network. We present extensive experiments and demonstrate that the proposed network obtains accuracies comparable to the state-of-the-art methods while significantly reducing the necessary memory space. Our best classification accuracy was 60.17% on the challenging composite dataset containing five objectives classes. Our method takes only 24.6 ms for classifying a ME video clip (less than the occurrence time of the shortest ME which lasts 40 ms). Our CNN design is suitable for real-time embedded applications with limited memory and computing resources.

Keywords: computer vision; deep learning; optical flow; micro facial expressions; real-time processing

1. Introduction

Emotion recognition has received much attention in the research community in recent years. Among the several sub-fields of emotion analysis, studies of facial expression recognition are particularly active [1–4]. Most of the affective computing methods in the literature apply the emotion model presented by Ekman [5] that reported seven basic expressions: anger, fear, surprise, sadness, disgust, contempt and happiness. Ekman developed the Facial Action Coding System (FACS) to describe the facial muscle movements according to the action units , i.e., the fundamental actions of individual muscles or groups of muscles that can be combined to represent each of the facial expressions. These facial expressions can thus be labeled by codes based on the observed facial movements rather than from subjective classifications of emotion.

In contrast to the traditional macro-expression, people are less familiar with micro facial expressions [5,6], and even fewer know how to capture and recognize them. A Micro-Expression (ME) is a rapid and involuntary facial expression that exposes a person's true emotion [7]. These subtle expressions usually take place when a person conceals his or her emotions in one of the two scenarios: conscious suppression or unconscious repression. Conscious suppression happens when one deliberately prevents oneself from expressing genuine emotions. On the contrary, unconscious repression occurs when the subject is not aware of his or her true emotions. In both cases, MEs reveal the subject's true emotions regardless of the subject's awareness. Intuitively, ME recognition has a vast number of potential applications across different sectors, such as the security field, neuromarketing [8], automobile drivers' monitoring [9] and lies and deceit detection [6].

Psychological research shows that facial MEs generally are transient (e.g., remaining less than 200 ms) and very subtle [10]. The short duration and subtlety levy great challenges on a human trying to perceive and recognize them. To enable better ME recognition by humans, Ekman and his team developed the ME Training Tool (METT). Even with the help of this training tool, human can barely achieve around 40% accuracy [11]. Moreover, humans' decisions are prone to being influenced by individual perceptions that vary among subjects and across time, resulting in less objective results. Therefore, a bias-free and high-quality automatic system for facial ME recognition is highly sought after.

A number of earlier solutions to automate facial ME recognition have been based on geometry or appearance feature extraction methods. Specifically, geometric-based features encode geometric information of the face, such as shapes and locations of facial landmarks. On the other hand, appearance-based features describe the skin textures of faces. Most existing methods [12,13] attempt to extract low-level features, such as the widely used Local Binary Pattern from Three Orthogonal Planes (LBP-TOP) [14–16] from different facial regions, and simply concatenate them for ME recognition. Nevertheless, transient and subtle ME inherently makes it challenging for low level-features to effectively capture essential movements in ME. At the same time, these features can also be affected by irrelevant information or noise in video clips, which further weakens their discrimination capabilities, especially for inactive facial regions that are less dynamic [17].

Recently, more approaches based on mid-level and high-level features have been proposed. Among these methods, the pipeline composed of optical flow and deep learning has demonstrated its high effectiveness for MEs recognition in comparison with traditional ones. The studies applying deep learning to tackle the ME classification problem usually considered well-known Convolutional Neural Networks (CNNs) such as ResNet [18] and VGG [19]. These studies re-purposed the use of off-the-shelf CNNs by giving them input data taken from the optical flow extracted from the MEs. While achieving good performance, these neural networks are quite demanding in terms of memory usage and computation.

In specific applications, for example, during automobile driver monitoring or student comprehension recognition in virtual education systems, fast and effective processing methods are necessary to capture emotional responses as quickly as possible. Meanwhile, thanks to great progresses in parallel computing, parallelized image processing devices such as embedded systems are easily accessible and affordable. Already well-adopted in diverse domains, these devices possess multiple strengths in terms of speed, embeddability, power consumption and flexibility. These advantages, however, are often at the cost of limited memory and computing power.

The objective of this work was to design an efficient and accurate ME recognition pipeline for embedded vision purposes. First of all, our design took into account thorough investigations on different CNN architectures. Next, different optical flow representations for CNN inputs were studied. Finally, our proposed pipeline achieved accuracy for ME recognition that is competitive with state-of-the-art approaches while being real-time capable and using less memory. The paper is organized as follows. In Section 2, several recent related studies are reviewed. Section 3 explains the proposed methodology in order to establish cost-effective CNNs for fast ME recognition. Section 4 provides experimental results and performance evaluations. Lastly, Section 5 concludes the paper.

2. Related Works

MEs begin at the onset (first frame where the muscles of the facial expressions start to contract), finish at the offset (last frame, where the face returns to its neutral state) and reach their pinnacle at the apex frames (see Figure 1). Because of their very short duration and low intensity, ME recognition and analysis are considered difficult tasks. Earlier studies proposed using low-level features such as LBP-TOP to address these problems. LBP-TOP is a 3D descriptor extended from the traditional 2D LBP. It encodes the binary patters between image pixels, and the temporal relationship between pixels and their neighboring frames. The resulting histograms are then concatenated to represent the temporal changes over entire videos. LBP-TOP has been widely adopted in several studies. Pfister et al. [14] applied LBP-TOP for spontaneous ME recognition. Yan et al. [15] achieved 63% ME recognition accuracy on their CASME II database using LBP-TOP. In addition, LBP-TOP has also been used to investigate differences between micro-facial movement sequences and neutral face sequences.

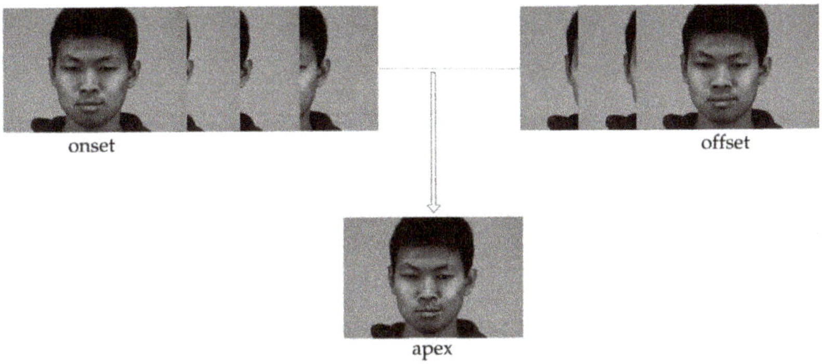

Figure 1. Example of a Micro-Expression (ME): the maximum movement intensity occurs at the apex frame.

Several studies aimed to extend low-level features extracted by LBP-TOP as they still could not reach satisfactory accuracy. For example, Liong et al. [20] proposed assigning different weights to local features, thereby putting more attention on active facial regions. Wang et al. [12] studied the correlation between color and emotions by extracting LBP-TOP from the Tensor-Independent Color Space (TICS). Ruiz-Hernandez and Pietikäinen [21] used the re-parameterization of second order Gaussian jet on the LBP-TOP, achieving a promising ME recognition result on the SMIC database [22]. Considering that LBP-TOP consists of redundant information, Wang et al. [23] proposed the LBP-Six Intersection Points (LBP-SIP) method which is computationally more efficient and achieves higher accuracy on the CASEME II database. We also note that the STCLQP (Spatio-Temporal Completed Local Quantization Patterns) proposed by Huang et al. [24] achieved a substantial improvement for analyzing facial MEs.

Over the years, as research showed that it is non-trivial for low-level features to effectively capture and encode a ME's subtle dynamic patterns (especially from inactivate regions), other methods shifted to exploit mid or high-level features. He et al. [17] developed a novel multi-task mid-level feature learning method to enhance the discrimination ability of the extracted low-level features. The mid-level feature representation is generated by learning a set of class-specific feature mappings. Better recognition performance has been obtained with more available information and features more suited to discrimination and generalization. A simple and efficient method known as Main Directional Mean Optical-flow (MDMO) was employed by Liu et al. [25]. They used optical flow to measure the subtle movement of facial Regions of Interest (ROIs) that were spotted based on the FACS. Oh et al. [26] also applied the monogenic Riesz wavelet representation in order to amplify subtle movements of MEs.

The aforementioned methods indicate that the majority of existing approaches heavily rely on hand-crafted features. Inherently, they are not easily transferable as the process of feature crafting and selection depends heavily on domain knowledge and researchers' experience. In addition, methods based on hand-crafted features are not accurate enough to be applied in practice. Therefore, high-level feature descriptors which better describe different MEs and can be automatically learned are desired. Recently, more and more vision-based tasks have shifted to deep CNN-based solutions due to their superior performance. Recent developments in ME recognition have also been inspired by these advancements by incorporating CNN models within the ME recognition framework.

Peng et al. [27] proposed a two-stream convolutional network DTSCNN (Dual Temporal Scale Convolutional Neural Network) to address two aspects: the overfitting problem caused by the small sizes of existing ME databases and the use of high-level features. We can observe four characteristics of the DTSCNN: (i) separate features were first extracted from ME clips from two shallow networks and then fused; (ii) data augmentation and higher drop-out ratio were applied in each network; (iii) two databases (CASME I and CASME II) were combined to train the network; (iv) the data fed to the networks were optical-flow images instead of raw RGB frames.

Khor et al. [28] studied two variants of an Enriched LRCN (Long-Term Recurrent Convolutional Network) model for ME recognition. Spatial Enrichment (SE) refers to channel-wise stacking of gray-scale and optical flow images as new inputs to CNN. On the other hand, Temporal Enrichment (TE) stacks obtained features. Their TE model achieves better accuracy on a single database, while the SE model is more robust against the cross-domain protocol involving more databases.

Liong et al. [29] designed a Shallow Triple Stream Three-dimentional CNN (STSTNet). The model takes input stacked optical flow images computed between the onset and apex frames (optical strain, horizontal and vertical flow fields), followed by three shallow Convolutional Layers in parallel and a fusion layer. The proposed method is able to extract rich features from MEs while being computationally light, as the fused features are compact yet discriminative.

Our objective was to realize a fast and high-performance ME recognition pipeline for embedded vision applications under several constraints, such as embeddability, limited memory and restricted computing resources. Inspired by existing works [27,29], we explored different CNN architectures and several optical flow representations for CNN inputs to find cost-effective neural network architectures that were capable of recognizing MEs in real-time.

3. Methodology

The studies applying deep learning to tackle the ME classification problem [30–33] usually used pretrained CNNs such as ResNet [18] and VGG [19] and applied transfer learning to obtain ME features. In our work, we first selected off-the-shelf ResNet18 because it provided the best trade-off between accuracy and speed on the challenging ImageNet classification and was recognized for its performance in transfer learning. ResNet [18] explicitly lets the stacked layers fit a residual mapping. Namely, the stacked non-linear layers are let to fit another mapping of $F(x) := H(x) - x$ where $H(x)$ is the desired underlying mapping and x the initial activations. The original mapping is recast into $F(x) + x$ by feedforward neural networks with shortcut connections. ResNet18 has 20 Convolutional Layers (CLs) (17 successive CLs and 3 branching ones). Residual links after each pair of successive convolutional units are used and the kernel size after each residual link is doubled. As ResNet18 is designed to extract features from RGB color images, it requires inputs to have 3 channels.

In order to accelerate processing speed in the deep learning domain, the main current trend in decreasing complexity of CNN is to reduce the number of parameters. For example, Hui et al. [34] proposed a very compact LiteFlowNet which is 30 times smaller in the model size and 1.36 times faster in the running speed in comparison with the state-of-the-art CNNs for optical flow estimation. In [35], Rieger et al. explored parameter-reduced residual networks on in-the-wild datasets, targeting real-time head pose estimation. They experimented with various ResNet architectures with a varying number of layers to handle different image sizes (including low-resolution images). The optimized ResNet achieved state-of-the-art accuracy with real-time speed.

It is well known that CNNs are created for specific problems and therefore over-calibrated when they are used in other contexts. ResNet18 was made for end-to-end object recognition: the dataset used for training had hundreds of thousands of images for each class and more than a thousand classes in total. Based on that: (i) An ME recognition study considers at most 5 classes, and the datasets of spontaneous MEs are scarce and contain far fewer samples, and (ii) optical flows are high-level features contrary to low-level color features and so require shallower networks; we have empirically reduced the architecture of ResNet18 by iteratively removing residual layers. This allowed us to assess the influence of the depth of the network on its classification capacities in our context and therefore to estimate the relevant calibration of the network.

Figure 2 illustrates the reduction protocol: at each step the last residual layer with two CLs is removed and the previous one is connected to the fully connected layer. Only networks with an odd number of CL are therefore proposed. The weights of all CNNs are pretrained using ImageNet. As highlighted in Table 1, the decrease in the number of CLs has a significant impact on the number of learnable parameters of the network, which directly affects the forward propagation time.

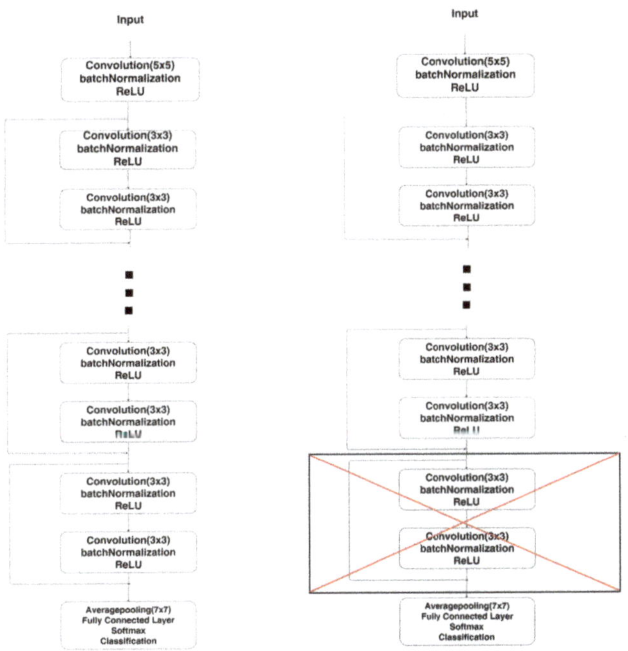

Figure 2. Depth reduction of a deep neural network: in the initial network, each residual layer contains two Convolutional Layers (CLs) (**left**); the last residual layer is removed (**middle**) to obtain a shallower network (**right**).

Table 1. Number of CLs and the number of learnable parameters in the proposed architectures.

CL	17	15	13	11	9	7	5	3	1
Nb. of param.	10,670,932	5,400,725	2,790,149	1,608,965	694,277	398,597	178,309	104,197	91,525

Once the network depth has been correctly estimated, the dimensions of the input have to be optimized. In our case, CNNs take optical flows extracted between the onset and apex frames of

ME video clips. It is between these two moments that the motion is most likely to be the strongest. The dimensionality of inputs determines the complexity of the network that uses them, since the reduction in input channels dictates the number of filters to be used throughout all following layers of the CNN. The optical flow between the onset (Figure 3a) and the apex (Figure 3b) typically has a 3-channel representation to be used in a pretrained architecture designed for 3-channel color images. This representation, however, may not be optimal for ME recognition.

From the assumption of brightness invariance, the movement of each pixel between frames over a period of time is estimated and represented as a vector (Figure 3c) indicating the direction and intensity of the motion. The projection of the vector on the horizontal axis corresponds to the Vx field (Figure 3d) while its projection on the vertical axis is the Vy field (Figure 3e). The Magnitude (M) is the norm of the vector (Figure 3f). Figure 4 illustrates this representation of one optical flow vector.

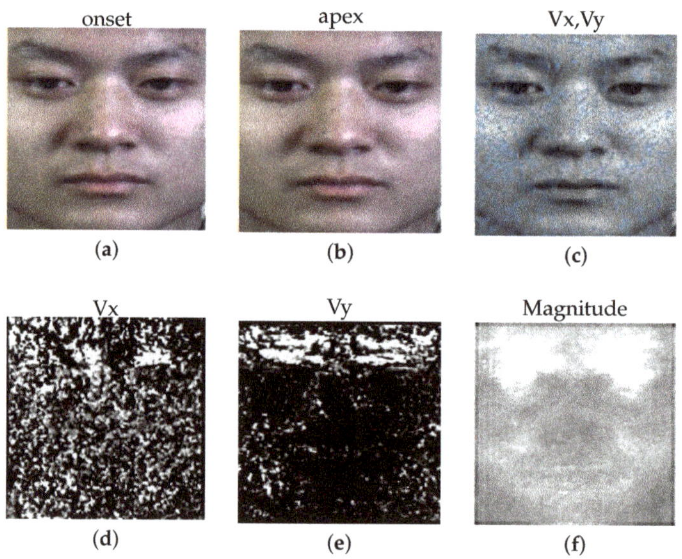

Figure 3. Optical flow is computed between the onset (**a**) and the apex (**b**): vectors obtained for a random sample of pixels (**c**), Vx field (**d**), Vy field (**e**) and Magnitude field (**f**).

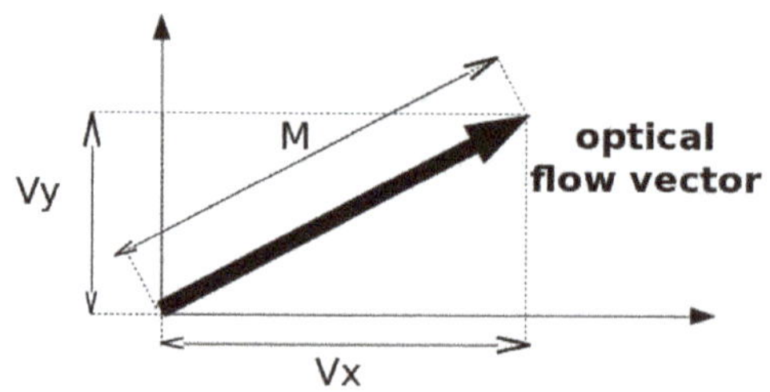

Figure 4. Visualization of M, Vx and Vy for one optical flow vector.

When classifying ME, the resulting matrices Vx , Vy and M are traditionally given as inputs to the CNN. Nonetheless, the third channel is inherently redundant since M is computed from Vx and Vy. Optical flow composed of the 2-channel Vx and Vy field could already provide all relevant information. Furthermore, we hypothesize that even a single channel motion field itself could be descriptive enough. Hence, we have created and evaluated networks taken as input for the optical flow in a two-channel representation (Vx-Vy) and in an one-channel representation (M, Vx or Vy).

For this purpose, the proposed networks begin with a number of CLs related to the depth optimization followed by a batch normalization and ReLU. Then the networks end with a maxpooling layer and a fully connected layer. The Figure 5 presents the architectures used with one to four CL according to the results of the experiments in Section 4. As illustrated in Table 2, a low dimensional input leads to a significant reduction in the number of learnable parameters and therefore in the complexity of the system.

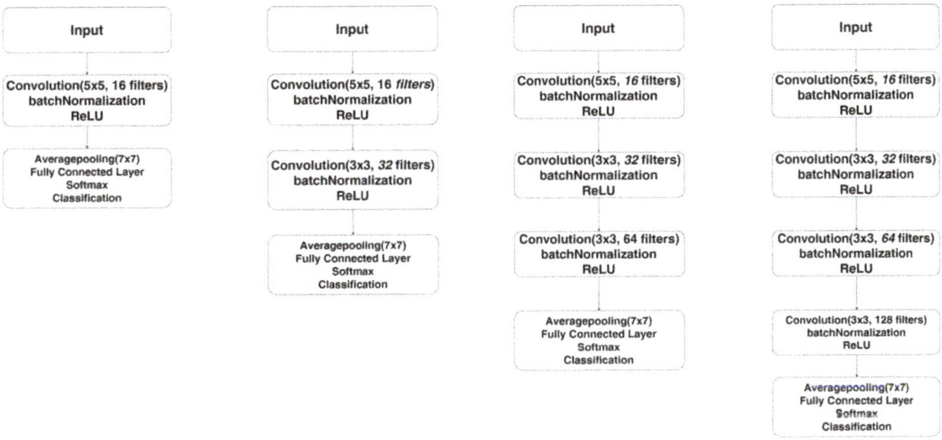

Figure 5. Proposed networks composed of one to four (from left to right) CLs for various representations of the optical flow as input.

Table 2. Number of learnable parameters according to the dimensionality of the input of the network.

Input	1 CL	2 CL	3 CL	4 CL
Single Channel	82,373	168,997	333,121	712,933
Double Channel	165,541	348,005	709,477	1,620,197

4. Experiments

4.1. Dataset and Validation Protocol Presentation

Two ME databases were used in our experiments. CASME II (Chinese Academy of Sciences Micro-Expression) [15] is a comprehensive spontaneous ME database containing 247 video samples, collected from 26 Asian participants with an average age of 22.03 years old. Compared to the first database, the Spontaneous Actions and Micro-Movements (SAMM) [36] is a more recent one consisting of 159 micro-movements (one video for each). These videos were collected spontaneously from a demographically diverse group of 32 participants with a mean age of 33.24 years old and a balanced gender split. Originally intended for investigating micro-facial movements, SAMM initially collected the seven basic emotions.

Both the CASME II and SAMM databases were recorded at a high-speed frame rate of 200 fps. They also both contain "objective classes," as provided in [37]. For this reason, the Facial MEs Grand Challenge 2018 [38] proposed to combine all samples from both databases into a single composite dataset of 253 videos with five emotion classes. It should be noted that the repartition is not very well balanced. Namely, this composite database is composed of 19.92% "happiness", 11.62% "surprise", 47.30% "anger", 11.20% "disgust" and 9.96% "sadness".

Similarly to [38], we applied the Leave One Subject Out (LOSO) cross-validation protocol for ME classification, wherein one subject's data is used as a test set in each fold of the cross-validation. This is done to better reproduce realistic scenarios where the encountered subjects are not present during training of the model. In all experiments, recognition performance is measured by accuracy, which is the percentage of correctly classified video samples out of the total number of samples in the database.

The Horn–Schunck method [39] was selected to compute optical flow. This algorithm was widely used for optical flow estimation in many recent studies for virtue of its robustness and efficiency. Throughout all experiments, we trained the CNN models with a mini-batch size of 64 for 150 epochs using the RMSprop optimization. Feature extraction and classification were both handled by the CNN. Simple data augmentation was applied to double the training size. Specifically, for each ME video clip used for training, in addition to the optical flow between the onset and apex frame, we also included a second flow computed between the onset and apex+1 frame.

4.2. ResNet Depth Study

In order to find the ResNet depth which permits an optimal compromise between the ME recognition performance and the number of learnable parameters, we tested different CNN depths using the method described in Section 3. The obtained accuracies are given in Table 3:

Table 3. Accuracies varied by the number of Convolutional Layers (CLs) and associated number of learnable parameters.

Nb. of CL	17	15	13	11	9	7	5	3	1
Nb. of param.	10,670,932	5,400,725	2,790,149	1,608,965	694,277	398,597	178,309	104,197	91,525
Accuracy	57.26%	57.26%	60.58%	59.34%	60.17%	61.00%	58.51%	60.17%	58.92%

We observed that the best score was achieved by ResNet8, which had seven CLs. However, the scores achieved by different numbers of CL did not vary much. Furthermore, beyond seven CL, adding more CL did not improve the accuracy of the model. The fact that accuracy does not increase along with depth confirms that multiple successive CL are not necessary to achieve a respectable accuracy. The most interesting observation was that with a single CL, we achieved a score that is not very far from the optimal score while the size of the model was much more concise. This suggests that instead of deep learning, a more "classical" approach exploiting shallow neural networks presents an interesting field to explore when considering portability and computational efficiency for embedded systems. That is the principal reason we restricted our study to shallow CNNs.

4.3. CNN Input Study

In this subsection, we study impacts of optical flow representations on ME recognition performance. Two types of CNN have been investigated, one with 1-channel input (Vx, Vy, or M) and the other one using the 2-channel Vx-Vy pair. Due to the fact that off-the-shelf CNNs typically take 3-channel inputs and are pretrained accordingly, applying transfer learning to adapt to our models would have been a nontrivial task. Instead, we created custom CNNs and trained them from scratch. Table 4 shows the recognition accuracies of different configurations using a small number of CNN layers.

Table 4. Accuracies under various CNN architectures and optical flow representations.

	1 CL	**2 CL**	**3 CL**	**4 CL**
Vx	52.24%	54.34%	53.92%	53.50%
Vy	58.09%	59.34%	**60.17%**	60.17%
Vx-Vy	58.51%	59.75%	**60.17%**	58.09%
M	58.09%	58.92%	59.34%	59.34%

We can observe that the Vx-Vy pair and Vy alone gave the best results, both representations achieving 60.17% accuracy. On the other hand, using Magnitude alone leads to a similar accuracy to those of Vy and the Vx-Vy pair with a score of 59.34%. Vx got the worst results overall, with a maximum score of 54.34%. This observation indicates that the most prominent features for ME classification might indeed be more dominant in vertical movement rather than the horizontal movement. This assumption is logical when thinking about the muscle movements happening in each known facial expression.

To better visualize the difference in the high-level features present in Vx, Vy and the Magnitude, we did an averaging on all the different samples according to their classes. The result can be seen in Figure 6. We observed that Vx exhibits a non-negligible quantity of noise. Magnitude and Vy, on the other hand, had clear regions of activity for each class. The regions of activity were aligned with the muscles responsible of each facial expression.

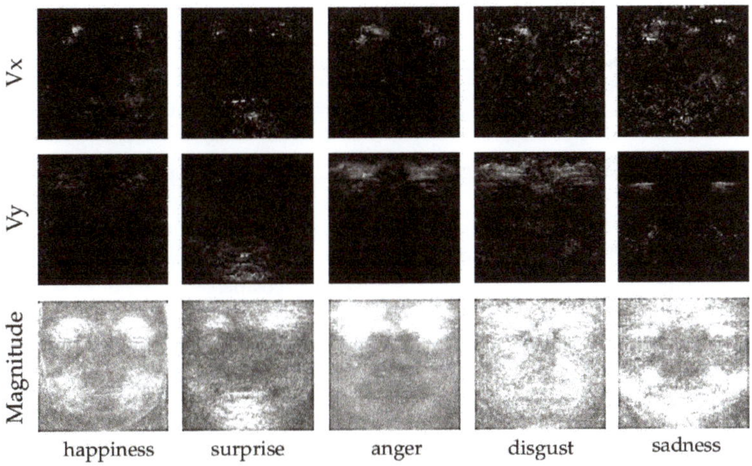

Figure 6. Average optical flow obtained in the dataset per ME class. Studied classes are in order from left to right: happiness, surprise, anger, disgust and sadness.

4.4. Classification Analysis

In order to understand obtained results, we measured cosine similarity of features extracted by three CNNs: ResNet8 (Section 4.2), Vx-Vy-3-CL and Vy-3-CL (Section 4.3). Usually, the convolutional layers of CNNs are considered as different feature extractors, only the last fully connected layer directly performs the classification task. The features just before classification can be represented in vector format. Cosine similarity measures the similarity between two vectors a and b using Equation (1):

$$cosine(a, b) = \frac{a^T b}{||a||\ ||b||} \tag{1}$$

Cosine similarity values fall within the range of $[-1, 1]$; values closer to 1 indicate higher similarity between two vectors. Tables 5–7 display the cosine similarity values: with two samples five ME classes, we calculated intra-similarity and average inter-similarity of each class using the same configuration for three CNNs.

Table 5. Cosine similarity for the 3-CL CNN with single-channel input Vy.

	Happiness	Surprise	Anger	Disgust	Sadness
Happiness	0.6007	0.1320	0.0574	0.0146	0.1154
Surprise	0.1320	0.5572	0.0485	0.0667	0.1415
Anger	0.0574	0.0485	0.5260	0.0318	0.0698
Disgust	0.0146	0.0667	0.0318	0.5663	0.0159
Sadness	0.1154	0.1415	0.0698	0.0159	0.5099

Table 6. Cosine similarity for the 3-CL CNN with double-channel inputs (Vx-Vy).

	Happiness	Surprise	Anger	Disgust	Sadness
Happiness	0.5615	0.1700	0.1171	0.1155	0.1195
Surprise	0.1700	0.5831	0.1432	0.1502	0.1618
Anger	0.1171	0.1432	0.5672	0.1176	0.1503
disgust	0.1155	0.1502	0.1176	0.5447	0.1225
Sadness	0.1195	0.1618	0.1503	0.1225	0.5443

Table 7. Cosine similarity for ResNet8.

	Happiness	Surprise	Anger	Disgust	Sadness
Happiness	0.8464	0.3966	0.3860	0.3126	0.2960
Surprise	0.3966	0.8159	0.4040	0.3362	0.3324
Anger	0.3860	0.4040	0.8344	0.3654	0.3307
Disgust	0.3126	0.3362	0.3654	0.8598	0.2363
Sadness	0.2960	0.3324	0.3307	0.2363	0.9343

Firstly, we observed that diagonal values (intra-class) across all three CNNs were significantly higher in comparison with other values (inter-class). This illustrates that all three CNNs are capable of separating different ME classes. Secondly, the intra-class cosine similarity of ResNet is closer to 1, suggesting that ResNet features are more discriminative. We hypothesize that our simplified CNNs with reduced layers extract less refined features, resulting in the minor decrease in performance (61.00% vs. 60.17%).

4.5. Performance Evaluations

In this subsection, we describe measuring our proposed method in three aspects: recognition accuracy, needed memory space and processing speed. Since we obtained optimal results by using the Vy field and 3-layer CNN, further evaluations concentrated on this particular configuration.

Evaluation on recognition accuracy: We performed an accuracy comparison of five objective ME class recognition (see Table 8). Our best CNN reached a similar performance as those of other studies using the same protocol of validation. It is worth mentioning that Peng et al. [40] employed a macro-to-micro transferred ResNet10 model and obtained a better result. Their work used four Macro-Expression datasets (>10 K images) and some preprocessing, such as color shift, rotation and smoothing. These additional operations make their proposed method difficult for deployment on embedded systems. After seeing the confusion matrix of our model (Figure 7), we also noticed that the distribution of correct assessments for Vy was more balanced than the ones gotten from [28] (Figure 8).

Table 8. Comparison between our method and those of other top-performers from literature.

Method	Accuracy
LBP_TOP [28]	42.29%
Khor et al. [28]	57.00%
Peng et al. [40]	74.70%
Proposed method	60.17%

The DTSCNN proposed by Peng et al. in [27] opted for two optical flows computed differently from a ME sample, which made the whole network robust to different frame rates of ME videos. In detail, the first optical flow is calculated using 64 frames around the apex to adapt to the frame rate of CASME I. Similarly, the second optical flow is given by the 128 frames around the apex adapted to the frame rate of CASME II. In case the number of frames composing the ME is not sufficient, a linear interpolation method is used to normalize the video clips. Their study used two CNNs in parallel to extract two separate features before concatenating them. The resultant feature vector was then fed as input to an SVM to be classified. The DTSCNN was tested on four classes (positive, negative, surprise and other) from a composite dataset consisting of the CASME I and CASME II databases, and it achieved an average recognition rate of 66.67%. The STSTNet proposed by Liong et al. in [29] makes use of three-dimensional CNNs which carry out three-dimensional convolutions instead of two-dimensional ones (such as ResNet, VGG, the networks presented in [27,28,40] and our study). It was tested on three classes: positive, negative and surprise from a composite database consisting of samples from the SMIC, CASME II and SAMM databases. It achieved an unweighted average recall rate of 76.05% and an unweighted F1-score of 73.53%. Both of these two frameworks are not very suitable for real-time embedded applications constrained by limited memory and computing resources.

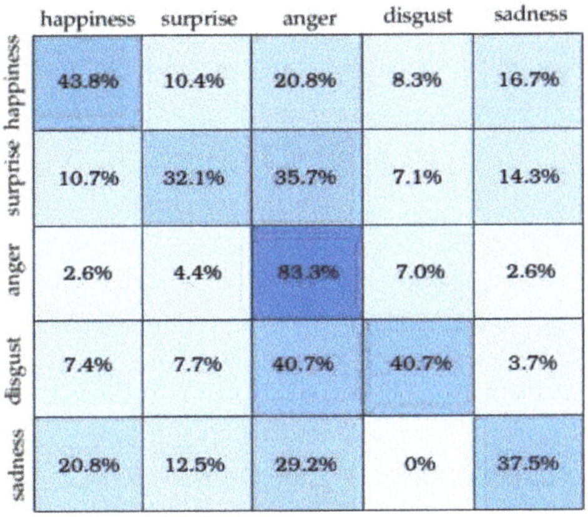

Figure 7. Confusion matrix corresponding to our network with 3 CLs and Vy as input.

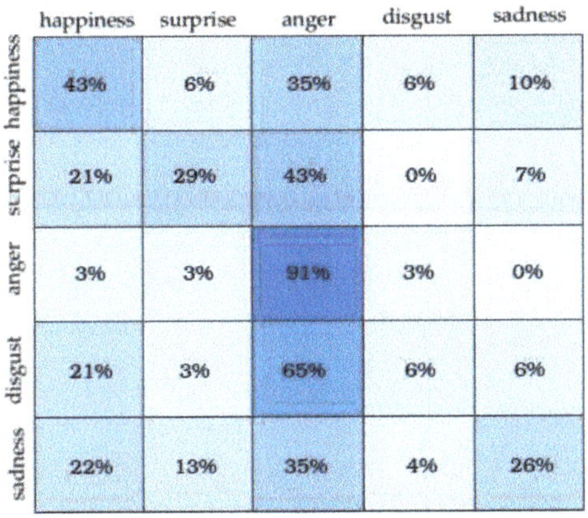

Figure 8. Confusion matrix obtained by the work of [28].

Evaluation on memory space: Table 9 summarizes the number of learnable parameters and used filters according to the dimensionality of the network inputs. The minimum required memory space corresponds to 333,121 parameter storage, which is less than 3.12% of that of off-the-shelf ResNet18.

Table 9. Number of learnable parameters and filters (in brackets) of various network architectures under different input dimensions.

Input	1 CL	2 CL	3 CL	4 CL
Single channel	82,373 (16)	168,997 (48)	333,121 (112)	712,933 (240)
Double channel	165,541 (32)	348,005 (96)	709,477 (224)	1,620,197 (480)

Evaluation on processing speed: We used a mid-range computer with an Intel Xeon processor and an Nvidia GTX 1060 graphic card to carry out all the experiments. The complete pipeline was implemented in MatLAB 2018a with its deep learning toolbox. Our model which achieved the best score was the CNN with a single-channel input and three successive CL. It needs 12.8 ms to classify the vertical component Vy. The optical flow between two frames requires 11.8 ms to compute using our computer, leading to a total runtime to classify an ME video clip of 24.6 ms. In our knowledge, the proposed method outperforms most ME recognition systems in terms of processing speed.

5. Conclusions and Future Works

In this paper, we propose cost-efficient CNN architectures to recognize spontaneous MEs. We first investigated the depth of the well-known ResNet18 network to demonstrate that using only a small number of layers is sufficient in our task. Based on this observation, we have experienced several representations of network input.

Following several previous studies, we fed CNNs with optical flow estimated from the onsets and apexes of MEs. Different flow representations (horizontal Vx, vertical Vy, Magnitude M and the Vx-Vy pair) have been tested and evaluated on a composite dataset (CASME II and SAMM) for recognition of five objective classes. The results obtained on the Vy input alone are more convincing. That was likely due to the fact that such an orientation is more suitable describing ME's motion and its variations between the different expression classes. Experimental results demonstrated that the proposed method can achieve similar recognition rate when compared with state-of-the-art approaches.

Finally, we obtained an accuracy of 60.17% with a light CNN design consisting of three CLs with single-channel inputs Vy. This configuration enables the number of learnable parameters to be reduced by a factor of 32 in comparison with the ResNet18. Moreover, we achieved a processing time of 24.6 ms which is shorter than MEs (40 ms). Our study opens up an interesting way to find the trade-off between speed and accuracy in ME recognition. While the results are encouraging, it should be noted that our method does not provide better accuracy than the ones described in the literature. Instead, a compromise has to be made between accuracy and processing time. By minimizing the computation, our proposed method manages to obtain accuracy comparable to the state-of-the-art systems while being compatible with the real-time constraints of embedded vision.

Several future works could further enhance both the speed and accuracy of our proposed ME recognition pipeline. These include more advanced data augmentation techniques to improve recognition performance. Moreover, new ways to automatically optimize the structure of a network to make it lighter have been presented recently. Other networks optimized for efficiency will also be explored. For example, MobileNet [41] uses depth-wise separable convolutions to build light weight CNN. ShuffleNet [42] uses pointwise group convolution to reduce computation complexity of 1×1 convolutions and channel shuffle to help the information flowing across feature channels. Our next step of exploration aims to analyze and integrate these new methodologies in our framework. Furthermore, we also hope to investigate new emotional machines while avoiding AI modeling errors and biases [43].

Author Contributions: Conceptualization, C.M. and F.Y.; formal analysis, R.B. and Y.L.; methodology, R.B., Y.L. and C.M.; software R.B. and Y.L.; supervision, C.M., D.G. and F.Y.; writing—review and editing, R.B., Y.L. and C.M., D.G. and F.Y.; writing—original draft R.B. and Y.L.; funding acquisition, D.G. All authors have read and agreed to the published version of the manuscript.

Funding: This work was supported by the H2020 Innovative Training Network (ITN) project ACHIEVE (H2020-MSCA-ITN-2017: agreement no. 765866).

Conflicts of Interest: The authors declare no conflict of interest.

Abbreviations

The following abbreviations are used in this manuscript:

ME	Micro Expression
CL	Convolutional Layer
M	Magnitude

References

1. Shan, C.; Gong, S.; McOwan, P. Facial expression recognition based on local binary patterns: A comprehensive study. *Image Vis. Comput.* **2018**, *27*, 803–816. [CrossRef]
2. Edwards, J.; Jackson, H.; Pattison, P. Emotion recognition via facial expression and affective prosody in schizophrenia: A methodological review. *Clin. Psychol. Rev.* **2002**, *22*, 789–832. [CrossRef]
3. Busso, C.; Deng, Z.; Yildirim, S.; Bulut, M.; Lee, C.; Kazemzadeh, A.; Lee, S.; Neumann, U.; Narayanan, S. Analysis of emotion recognition using facial expressions, speech and multimodal information. In Proceedings of the 6th International Conference on Multimodal Interfaces, College, PA, USA, 13–15 October 2004; pp. 205–211.
4. Biondi, G.; Franzoni, V.; Gervasi, O.; Perri, D. An Approach for Improving Automatic Mouth Emotion Recognition. In *Lecture Notes in Computer Science, 11619 LNCS*; Springer, Cham: Switzerland, 2019; pp. 649–664.
5. Ekman, P.; Friesen, W.V. Nonverbal Leakage and Clues to Deception. *Psychiatry* **1969**, *32*, 88–106. [CrossRef] [PubMed]
6. Ekman, P. *Telling Lies: Clues to Deceit in the Marketplace, Politics, and Marriage (Revised Edition)*; WW Norton & Company: New York, NY, USA, 2009.
7. Haggard, E.; Isaacs, K. Micromomentary facial expressions as indicators of ego mechanisms in psychotherapy. In *Methods of Research in Psychotherapy*; Springer: Boston, MA, USA, 1966; pp. 154–165.
8. Vecchiato, G.; Astolfi, L.; Fallani, F. On the use of EEG or MEG brain imaging tools in neuromarketing research. *Comput. Intell. Neurosci.* 2011. [CrossRef] [PubMed]
9. Nass, C.; Jonsson, M.; Harris, H.; Reaves, B.; Endo, J.; Brave, S.; Takayama, L. Improving Automotive Safety by Pairing Driver Emotion and Car Voice Emotion. In Proceedings of the Extended Abstracts on Human Factors in Computing Systems; Association for Computing Machinery: New York, NY, USA, 2005; pp. 1973–1976.
10. Ekman, P. Lie Catching and Micro Expressions. In *The Philosophy of Deception*; Oxford University Press : Oxford, UK, 2009; pp. 118–133.
11. Frank, M.; Herbasz, M.; Sinuk, K.; Keller, A.; Nolan, C. I see how you feel: training laypeople and professionals to recognize fleeting emotions. In Proceedings of the Annual Meeting of International Communication Association, Chicago, IL USA, 21–25 May 2009.
12. Wang, S.J.; Yan, W.J.; Li, X.; Zhao, G.; Zhou, C.G.; Fu, X.; Yang, M.; Tao, J. Micro expression recognition using color spaces. *Trans. Image Process.* **2015**, *24*, 6034–6047. [CrossRef] [PubMed]
13. Wu, Q.; Shen, X.; Fu, X. The Machine Knows What You Are Hiding: An Automatic Micro-Expression Recognition System. In Proceedings of the Affective Computing and Intelligent Interaction, Memphis, TN, USA, 9–12 October 2011; pp. 152–162.
14. Pfister, T.; Li, X.; Zhao, G.; Pietikäinen, M. Recognising spontaneous facial micro- expressions. In Proceedings of the 2011 International Conference on Computer Vision, Barcelona, Spain, 6–13 November 2011; pp. 1449–1456.
15. Yan, W.; Li, X.; Wang, S.; Zhao, G.; Liu, Y.; Chen, Y.; Fu, X. CASMEII: An improved spontaneous micro-expression database and the baseline evaluation. *PLoS ONE* **2014**, *9*, e86041.

16. Davison, A.; Yap, M.; Costen, N.; Tan, K.; Lansley, C.; Leightley, D. Micro-facial movements: an investigation on spatio-temporal descriptors. In Proceedings of the European Conference on Computer Vision, Zürich, Switzerland, 6–12 September 2014; pp. 111–123.
17. He, J.; Hu, J.F.; Lu, X.; Zheng, W.S. Multi-task mid-level feature learning for micro-expression recognition. *Pattern Recognit.* **2017**, *66*, 44–52. [CrossRef]
18. He, K.; Zhang, X.; Ren, S.; Sun, J. Deep residual learning for image recognition. In Proceedings of the Conference on Computer Vision and Pattern Recognition, Las Vegas, NV, USA, 27–30 June 2016; pp. 770–778.
19. Simonyan, K.; Zisserman, A. Very deep convolutional networks for large-scale image recognition. In Proceedings of the International Conference on Learning Representations, San Diego, CA, USA, 7–9 May 2015.
20. Liong, S.T.; See, J.; Phan, R.W.; Ngo, A.L.; Oh, Y.H.; Wong, K. Subtle expression recognition using optical strain weighted features. In Proceedings of the Asian Conference on Computer Vision, Singapore, 1–5 November 2014.
21. Ruiz-Hernandez, J.; Pietikäinen, M. Encoding local binary patterns using re-parameterization of the second order Gaussian jet. In Proceedings of the International Conference on Automatic Face and Gesture Recognition, Shanghai, China, 22–26 April 2013; pp. 1–6.
22. Li, X.; Pfister, T.; Huang, X.; Zhao, G.; Pietikäinen, M. A spontaneous micro-expression database: Inducement, collection and baseline. In Proceedings of the International Conference on Automatic Face and Gesture Recognition, Shanghai, China, 22–26 April 2013; pp. 1–6.
23. Wang, Y.; See, J.; Phan, R.; Oh, Y. LBP with six intersection points: Reducing redundant information in LBP-TOP for micro-expression recognition. In Proceedings of the Asian Conference on Computer Vision, Singapore, 1–5 November 2014; pp. 525–537.
24. Huang, X.; Zhao, G.; Hong, X.; Zheng, W.; Pietikäinen, M. Spontaneous facial micro-expression analysis using Spatiotemporal Completed Local Quantized Patterns. *Neurocomputing* **2016**, *175*, 564–578. [CrossRef]
25. Liu, Y.J.; Zhang, J.K.; Yan, W.J.; Wang, S.J.; Zhao, G.; Fu, X. A Main directional mean optical flow feature for spontaneous micro-expression recognition. *Trans. Affect. Comput.* **2015**, *7*, 299–310. [CrossRef]
26. Oh, Y.H.; Ngo, A.C.L.; See, J.; Liong, S.T.; Phan, R.C.W.; Ling, H.C. Monogenic Riesz wavelet representation for micro-expression recognition. In Proceedings of the 2015 IEEE International Conference on Digital Signal Processing (DSP), Singapore, 21–24 July 2015; pp. 1237–1241.
27. Min, P.; Chongyang, W.; Tong, C.; Guangyuan, L.; Xiaolan, F. Dual Temporal Scale Convolutional Neural Network for Micro-Expression Recognition. *Front. Psychol.* **2017**, *8*, 1745–1757.
28. Khor, H.Q.; See, J.; Phan, R.C.W.; Lin, W. Enriched Long-term Recurrent Convolutional Network for Facial Micro-Expression Recognition. In Proceedings of the International Conference on Automatic Face & Gesture Recognition (FG 2018), Xi'an, China, 15–19 May 2018; Volume 1, pp. 667–674.
29. Liong, S.T.; Gan, Y.; See, J.; Khor, H.Q.; Huang, Y.C. Shallow Triple Stream Three-dimensional CNN (STSTNet) for Micro-expression Recognition. In Proceedings of the IEEE International Conference on Automatic Face & Gesture Recognition (FG 2019), Lille, France, 14–18 May 2019; pp. 1–5.
30. Patel, D.; Hong, X.; Zhao, G. Selective deep features for micro expression recognition. In Proceedings of the International Conference on Pattern Recognition (ICPR), Cancun, Mexico, 4–8 December 2016; pp. 2258–2263.
31. Li, Y.; Huang, X.; Zhao, G. Can micro-expression be recognized based on single apex frame? In Proceedings of the International Conference on Image Processing (ICIP), Athens, Greece, 7–10 October 2018; pp. 3094–3098.
32. Wang, S.J.; Li, B.J.; Liu, Y.J.; Yan, W.J.; Ou, X.; Huang, X.; Xu, F.; Fu., X. Micro-expression recognition with small sample size by transferring long-term convolutional neural network. *Neurocomputing* **2018**, *312*, 251–262. [CrossRef]
33. Gan, Y.; Liong, S.T.; Yau, W.C.; Huang, Y.C.; Tan., L.K. Off-apexnet on micro-expression recognition system. *Signal Proc. Image Comm.* **2019**, *74*, 129–139. [CrossRef]
34. Hui, T.W.; Tang, X.; Loy, C.C. LiteFlowNet: A Lightweight Convolutional Neural Network for Optical Flow Estimation. In Proceedings of the Computer Vision and Pattern Recognition, Salt Lake City, UT, USA, 18–23 June 2018.
35. Rieger, I.; Hauenstein, T.; Hettenkofer, S.; Garbas, J.U. Towards Real-Time Head Pose Estimation: Exploring Parameter-Reduced Residual Networks on In-the-wild Datasets. In Proceedings of the International Conference on Industrial, Engineering and Other Applications of Applied Intelligent Systems, Graz, Austria, 9–11 July 2019; pp. 122–134.

36. Davison, A.K.; Lansley, C.; Costen, N.; Tan, K.; Yap, M.H. SAMM: A Spontaneous Micro-Facial Movement Dataset. *Trans. Affective Comp.* **2018**, *9*, 116–129. [CrossRef]
37. Davison, A.K.; Merghani, W.; Yap, M.H. Objective classes for micro-facial expression recognition. *J. Imaging* **2018**, *4*, 119. [CrossRef]
38. Yap, M.H.; See, J.; Hong, X.; Wang, S.J. Facial Micro-Expressions Grand Challenge 2018 Summary. In Proceedings of the International Conference on Automatic Face & Gesture Recognition (FG 2018), Xi'an, China, 15–19 May 2018; pp. 675–678.
39. Horn, B.; Schunck, B. Determining optical flow. *Artif. Intell.* **1981**, *17*, 185–203. [CrossRef]
40. Peng, M.; Wu, Z.; Zhang, Z.; Chen, T. From Macro to Micro Expression Recognition: Deep Learning on Small Datasets Using Transfer Learning. In Proceedings of the International Conference on Automatic Face & Gesture Recognition (FG 2018), Xi'an, China, 15–19 May 2018; pp. 657–661.
41. Howard, A.G.; Zhu, M.; Chen, B.; Kalenichenko, D.; Wang, W.; Andreetto, T.W.M.; Adam, H. MobileNets: Efficient Convolutional Neural Networks for Mobile Vision Applications. *arXiv* **2017**, arXiv:1704.04861.
42. Zhang, X.; Zhou, X.; Lin, M.; Sun, J. ShuffleNet: An Extremely Efficient Convolutional Neural Network for Mobile Devices. In Proceedings of the Conference on Computer Vision and Pattern Recognition, Salt Lake City, UT, USA, 18–23 June 2018; pp. 6848–6856.
43. Vallverdù, J.; Franzoni, V. Errors, biases and overconfidence in artificial emotional modeling. In Proceedings of the IEEE/WIC/ACM International Conference on Web Intelligence Workshops, Thessaloniki, Greece, 14–17 October 2019; pp. 86–90.

applied sciences

MDPI

Review

Thermal Infrared Imaging-Based Affective Computing and Its Application to Facilitate Human Robot Interaction: A Review

Chiara Filippini [1,2,*], David Perpetuini [1], Daniela Cardone [1], Antonio Maria Chiarelli [1] and Arcangelo Merla [1,2]

1 Department of Neuroscience and Imaging, Institute for Advanced Biomedical Technologies, University G. D'Annunzio of Chieti-Pescara, Via Luigi Polacchi 13, 66100 Chieti, Italy; david.perpetuini@unich.it (D.P.); d.cardone@unich.it (D.C.); antonio.chiarelli@unich.it (A.M.C.); arcangelo.merla@unich.it (A.M.)
2 Next2U s.r.l., Via dei Peligni 137, 65127 Pescara, Italy
* Correspondence: chiara.filippini@unich.it; Tel.: +39-0871-3556954

Received: 6 March 2020; Accepted: 21 April 2020; Published: 23 April 2020

check for updates

Abstract: Over recent years, robots are increasingly being employed in several aspects of modern society. Among others, social robots have the potential to benefit education, healthcare, and tourism. To achieve this purpose, robots should be able to engage humans, recognize users' emotions, and to some extent properly react and "behave" in a natural interaction. Most robotics applications primarily use visual information for emotion recognition, which is often based on facial expressions. However, the display of emotional states through facial expression is inherently a voluntary controlled process that is typical of human–human interaction. In fact, humans have not yet learned to use this channel when communicating with a robotic technology. Hence, there is an urgent need to exploit emotion information channels not directly controlled by humans, such as those that can be ascribed to physiological modulations. Thermal infrared imaging-based affective computing has the potential to be the solution to such an issue. It is a validated technology that allows the non-obtrusive monitoring of physiological parameters and from which it might be possible to infer affective states. This review is aimed to outline the advantages and the current research challenges of thermal imaging-based affective computing for human–robot interaction.

Keywords: human–robot interaction; thermal IR imaging; affective computing; social robots; emotion recognition

1. Introduction

Human–robot interaction (HRI) can vary depending on the specific robotic function. In fact, robots have been effectively used in the last decade in therapy and educational interventions, but they have been also used where human interaction is not required, such as for example in robotic vacuum cleaner applications. Even though the latter were not designed to interact socially and form bonds with humans, it was found that people often attribute human-like characteristics to their robotic technology and that they express attachment toward them [1]. This is in line with the assertion that humans have a natural tendency to anthropomorphize everything around them, including technology [2]. This goes especially with children, who "are unlikely to only use a robot as a tool, and they will undoubtedly have some sort of interaction that can be considered social" [3]. Therefore, it would be desirable that robots, which are designed to interact with adults and children, would be able to socially interact. To facilitate the interaction, robots should be easy to use, and they should be built to understand and correctly respond to different needs. As robots gain utility, and thereby influence on society, research in HRI is becoming increasingly important. HRI primarily deals with social robotics and includes relevant aspects of robot action, perception, and cognition. The development of social robots focuses on the design of living machines that humans would perceive as realistic, effective, communicative, and cooperative [4]. For this purpose, social robots should be able to express, through their shapes and behaviors, a certain degree of "intelligence" [5]. This skill entails the whole set of social and perceptual abilities of the robot, delivering a human-like interaction.

However, on the other hand, the matter about the anthropomorphism of social robots is still strongly debated. Indeed, the inclination to anthropomorphize the existing technology could have negative consequences [6] and even lead to an unrecognized change in human relationships [7] or invasion of privacy [8]. Therefore, it would be prudent to distinguish cases in which the use of anthropomorphism can be encouraged and cases in which it would be better not to [9]. Darling [10] argues that anthropomorphic framing is desirable where it enhances the function of the technology. Especially, it should be encouraged for social robots and discouraged for robots that are not designed in an intrinsically social way [9,10]. The present review is focused on social robots and in this context, their anthropomorphism is aimed at facilitating the interaction with humans.

An important key to reproduce a human-like behavior and facilitate HRI is the understanding of human emotional reactions. In recent years, a great effort was devoted to endowing the robot with the capability of interpreting and adapting to humans' emotional state. Consequently, a core aspect of HRI is Affective Computing. The term Affective Computing was created over 10 years ago; it is defined as "computing that relates to, arises from, or deliberately influences emotion or other affective phenomena" [11]. Since then, the community developed different concepts, models, frameworks, and demo applications for affective systems that are able to recognize, interpret, process, and simulate human feelings and emotions. However, these activities have been currently limited to possibility studies, proof-of-concept prototypes, or demonstrators.

Human emotions are indeed manifested as visible changes in facial expressions, gestures, body movements, or voice tone [12]. Beyond these observable expressive channels, physiological modulations also occur and can be observed as modifications in blood pressure, heart rate, or electrodermal activity [13,14]. These modulations are controlled by the autonomic nervous system and are not directly visible to humans. The main approaches to objectively measure these emotional signs rely on the observation of the face, gestures or body posture, and on measurements of physiological parameters through contact sensors [15,16]. Visual observation has the advantage of being completely non-invasive and only moderately intrusive. In fact, whereas a camera is being considered an intrusion into privacy, people are inclined to accept and forget it once they get used to it and see a value in it [17]. Furthermore, the greatest advantage of using the visual domain for emotion recognition is that it relies on years of study and cutting-edge algorithms developed for face identification and facial expressions analysis. The disadvantage of the visual approach is that people are prone to avoid facial expressions when interacting with technical systems [18]. In fact, the facial display of emotional states characterizes the human-to-human interaction, and it was acquired through evolution to facilitate communication and to influence others' actions [19,20]. Since machines do not still respond to emotional expressions, humans have not yet learned to use this channel when communicating with a robot. On the contrary, measuring physiological parameters has the advantage of evaluating features that people cannot control or mask [21]. Therefore, they may deliver much more reliable data about the emotional symptoms than visual channels. Physiological parameters such as heart rate, skin temperature, and electrodermal activity are very simple to acquire and analyze. The drawback of working with physiological readings is that they are mostly obtained through contact sensors [22]. Direct contact with the person's skin requires the willingness to correctly wear the device. Novel wearable sensors have to meet various technological requirements (e.g., reliability, robustness, availability, and quality of data), which are often very difficult to obtain. Finally, the time required for sensors placing is not negligible.

In recent years, thermal InfraRed (IR) imaging has been used for Affective Computing [23–25], and it exploits the advantages of both visual and physiology measuring approaches, as well as overcoming their drawbacks. Thermal IR imaging-based Affective Computing is a breakthrough technology that enables monitoring human physiological parameters and Autonomic Nervous System (ANS) activity in a non-contact manner and without subject's constraining [26]. Although thermal IR imaging has been widely used for human emotion recognitions [27–35], its application in HRI implies overcoming ambitious and entirely new challenges. The most important ones are real-time monitoring and real-life applications. To this end, the use of a mobile and ideally miniaturized technology is essential to bridge the gap between laboratory and real-world applications. In addition, since robots should be commercially available, they need to include low-cost technology, which implies facing further issues such as low resolution and signal-to-noise ratio.

In this survey, the importance of facilitating HRI in different aspect of everyday life is firstly highlighted. The state-of-the-art of affective state recognition through thermal IR imaging is briefly reviewed. Afterwards, the above-mentioned challenges are examined, indicating what has been done so far and suggesting further development to ensure its future efficient use in the field. Finally, the possible impact of this technology for HRI applications in infants, children, and adults is highlighted.

2. Study Organization and Search Processing Method

This study has been carried out as a systematic literature review based on the original guidelines as proposed by Kitchenham [36]. The work aims to highlight the positive impact that the thermal IR imaging technology can have on HRI application, facilitating and enhancing this interaction by recognizing the human affective state. For this purpose, the research questions (RQ) addressed by this study were:

RQ1. What is the broad social impact of a facilitated HRI? Or Why is facilitated HRI important?

RQ2. What are the scientific bases for thermal IR imaging as an affective state recognition technique?

RQ3. What are the limitations of current research?

RQ4. Has thermal IR imaging already been addressed in the HRI field?

One of the goals of this study was to make this review as inclusive and practical as possible. Therefore, the databases searched were both Scopus and Google Scholar. All the papers published in conferences and journals between 2000 and 2020 were considered. Papers published from the year 2000 were considered, since the word "Affective Computing" was coined by Picard that year [6], and "Affective Computing" started to be applied in the HRI field. For each research question, a search process was applied.

Concerning RQ1, the search was based on the words "facilitate" and "human–robot Interaction". In the Scopus database, the survey was set up by searching for those words within the following fields: article title, abstract, and keywords. The basic search generated 441 results. In the Scholar database, on the other hand, the advanced search can be performed either by searching (i) the entire text or (ii) the title only. Therefore, the search was based on "facilitate human–robot interaction" within the entire text. A total of 360 results were obtained from the Scholar survey. The review papers and all the papers that did not refer to a user study or that did not relate to HRI were excluded, which reduced the considered pool to 155 papers. Within those papers, 130 were related to Scopus research, and 105 were from Scholar with an overlap of 80 papers. The manual review process was adopted for the final exclusion; the papers' abstracts were scanned, and all the papers that did not report experimental applications of human interaction were discarded. A total of 18 papers were discussed in the review concerning these keywords. Among those papers, 13 papers resulted from both Scopus and Scholar research, while 5 were from Scholar only.

With respect to RQ2 and RQ3, the searched keywords were "thermal imaging" OR "IR imaging" OR "thermography" AND "emotion recognition" OR "Affective Computing" OR "emotion". In the Scopus database, those keywords were surveyed in fields such as article title, abstract, and keywords. Whereas in Scholar, the advanced survey was carried out by searching for "thermal imaging" OR "IR imaging" OR "thermography" with at least one of these words: "Affective Computing" OR "emotion", and the field searched was the entire text. The search generated 115 results in Scopus and 163 in Scholar with an overlap of 95 papers. The results were scanned through a manual review procedure focused on the papers' abstracts, which aimed to identify whether the considered works reported on thermal IR imaging for human emotion recognition. Papers not related to it were excluded. The resulting papers were analyzed and grouped based on their experimental applications, strengths and limitations were indicated, and 46 were reported in the present work. Among those papers, 40 papers resulted from both Scopus and Scholar research, while 6 were from Scholar only.

Finally, as for the RQ4, the keywords "thermal imaging" OR "IR imaging" OR "thermography" AND "human–robot interaction" were searched. The fields checked and the procedure performed were the same reported in the previous RQs. The basic search generated 13 results in Scopus and 388 in Scholar. Ten papers that resulted from Scopus search were also found in the Scholar research outcome. Exclusion criteria regarded all the results that were not conference or journal papers actually related to thermal IR imaging applied in the HRI field. From the latter, 20 papers of thermal IR imaging-based Affective Computing were selected and included in this work. Among those papers, a subset of 2 papers was linked to both RQ2/RQ3 and RQ4; therefore, it was reported in both sections.

3. The Importance of Facilitating Human–Robot Interaction

Robotic technologies and HRI are being increasingly integrated in real-life contexts. Modern society creates ever more spaces where robot technology is intended to interact with people. Robots applications can range from education to communication, assistance, entertainment, healthcare, and tourism. Hence, there is a need to better understand how robotic technologies shape the social contexts in which they are used [37]. In this section, the importance of a natural and efficient HRI, in three major fields such as education, healthcare, and tourism, is deepened.

Economic and demographic factors drive the need for technological support in education. The growing number of students per class, the reduction of the school budget, and the demand for greater personalization of curricula for children with diverse needs are encouraging research on technology-based support for parents and teachers [38]. Hence, the efficacy of robots in education is of primary interest. Movellan et al. deployed a fully autonomous social robot in a nursery school's classrooms for a period of 2 weeks in order to study whether the robot could improve target vocabulary skills in toddlers (18–24 months age) [39]. The results showed that the vocabulary skills improved significantly; in fact, the number of learned words increased by 27% when compared to a matched set of control. Considerable educational benefits can also be obtained from a robot that takes on the role of a novice (i.e., a care-receiving robot), thus allowing the student to take on the role of the instructor, which generally improves self-confidence and learning outcomes. This phenomenon is known as learning by teaching. An example is the educational use of Pepper (Figure 1), which was designed to learn together with children at their home environment from a remote teacher [40].

Figure 1. Pepper robot interaction with a child.

Social robots are also widely used in healthcare fields. Given the exponential growth of vulnerable populations (e.g., the elderly, children with developmental disabilities, and sick people), there is an increasing demand for social robots that are able to provide aids and entertainment for patients in the hospital or at home. For instance, an important contribution can be provided by companion robots especially among sick people, in the mitigation of boredom, depression, isolation, and loneliness. In this perspective, Banks et al. explored the ability of a robotic dog (AIBO) to treat loneliness in elderly patients living in long-term care facilities (LTCF). Results demonstrated that LTCF residents showed a high level of attachment to AIBO, highlighting the capability of interactive robotic dogs to reduce loneliness [41].

Together with companion robots, therapy robots are also considered of high social impact. Therapy robots are generally employed to deliver treatment for people with physical and mental diseases, such as Autism Spectrum Disorder (ASD). A recent review reported statistics showing an annual increase in the number of children with ASD, starting from 1 out of 1000 children in 1970 up to 1 out of 59 children in 2018 [42]. Thus, the need for innovative care and proper attention for ASD children is compelling. Particularly, researches have shown that people suffering from Autism Spectrum Disorders (ASD) responded better to treatments involving robotic technology rather than treatments from human therapists [43]. Moreover, it was demonstrated that the use of robots in education helps ASD children to improve their abilities to handle social and sensory challenges at school environment and to better control the anxiety and stress [44,45]. Furthermore, Wilson et al. noted that one of the major barriers to the education of children with autism pertains to lack of knowledge, training, and specialized support staff as well as the lack of adequate resources for education and classroom size [46]. Therefore, the development of robots with advanced emotion recognition and HRI capabilities that can be used at school or at home is becoming essential in supporting ASD patients.

Finally, several applications in the literature confirmed the importance of HRI in tourism settings. In fact, with recent technological advancements in artificial intelligence (AI) and robotics, we see an increasing number of service robots entering tourism and hospitality contexts, including consumer-facing ones [47]. For instance, Niculescu et al. developed SARA (Singapore's Automated Responsive Assistant), a robotic virtual agent, to offer information and assistance to tourists, being able to detect the user's location on a map [48]. CLARA is a virtual restaurant recommendation system and conversational agent that provides tourists with information about sightseeing, restaurants, transportation, and general information about Singapore [49]. However, the adoption of service robots inevitably changes the nature of service experience. Unlike industrial robots whose performance metrics depend entirely on efficiency, the success of service robots depends on user satisfaction and, consequently on the degree of empathy and natural interaction. Tussyadiah et al. focused on consumer's evaluation of hotel service robots. In their study, they surveyed consumer response to two types of robots, NAO and Relay. The results revealed that consumer intention to accept hotel service robots is influenced by their interaction with the robot, dimensions of anthropomorphism, perceived intelligence and security [47]. The same conclusion was drawn from Kervenoael et al. in [50], where they stated that empathy from service robots has a positive and significant effect on the intention to use a robot. Empathy is meant as the robot's ability to understand or feel what another person is experiencing from within their frame of reference, i.e., the robot's ability to recognize emotion and to respond it in an appropriate way. In conclusion, beyond the requirement of well-programmed social robots, in order to cater to specific consumers or interlocutor's needs, they must incorporate a seamless integration of safe and reliable service that includes courtesy and inspires trust.

With sophisticated robots or artificial agents becoming ever more ubiquitous in daily life, their appropriate use is crucial. In this section, we provided examples of their positive impact on fields such as education, healthcare, and tourism. Research on HRI requires contributions and expertise from heterogeneous disciplines, including engineering and artificial intelligence. In fact, to ensure a natural HRI, social robots need to recognize human emotions, respond adequately to them, understand and generate natural language, have reasoning skills, plan actions, and execute movements in line with what is required by the specific context or situation [51]. A good extent of the effort can be ascribed to emotion recognition, which currently requires exploring fields of facial expression analysis and speech processing, which are anything but trivial tasks. Relying on emotion recognition through non-obtrusive physiological sensing, thermal IR imaging-based affective computing can be a way to avoid some of those challenges. Although this technique is already used in the field of emotion recognition, methodological considerations are required to make it suitable for HRI applications.

4. Affective States Recognition through Thermal IR Imaging

A considerable number of studies have explored the use of thermal IR imaging for classifying affective states and human emotions. Those studies were based on measuring a person's physiological cues that are considered related to affective states. Indeed, the observations of affective nature derive primarily from muscular activity, subcutaneous blood flow, perspiration patterns in specific body parts, as well as from changing in breathing or heart rate, which are all phenomena that are controlled by our ANS. Measuring facial cutaneous temperature and assessing both its topographic and temporal distribution can provide insights about the person's autonomic activity. This depends on the ANS's role in the human body's thermal homeostasis and in the regulation of physiological responses to emotional stimuli [29]. Alert, anxiety, frustration responses, and other affective states determine the redistribution of the blood in the vessels through vasodilation or vasoconstriction phenomena which are regulated by ANS. These phenomena can be captured by an IR thermal camera through changes in the IR emissivity of the skin. Vasoconstriction implies a decrease in temperatures of the Region Of Interest (ROI). On the other hand, vasodilatation is the cause of heating. Vasomotor processes can be identified and monitored over time because they produce a thermal variation of the skin and they can be characterized by simple metrics such as temperature difference between data at two temporal points. For this reason, most researchers have focused on studying the relationship between thermal directional changes (i.e., temperature drop and rise) of specific skin areas in relation to psychophysiological states [28,31,52,53]. As a body area of interest, the human face is considered of particular importance since it can be easily recorded and it is naturally exposed to social stimuli. Regions of the face extensively characterized are the nose or nose tip, the glabella (area associated with the corrugator muscle), the periorbital area, forehead, and the orbicularis oculi (surrounding the eyes), as well as the maxillary area or the upper lip (perinasal) [24,28,54]. Partially evaluated regions were cheeks, carotid, eyes, fingers, and lips. An exhaustive review on this topic by Ioannou et al. summarized the emotions, the observed regions, and the direction of the average temperature changes in those regions (Table 1) [55].

Table 1. Overview of the temperature variations direction, as estimated through thermal IR imaging, for each emotion in the different regions considered. The table is adapted from [55].

	Stress	Fear	Startle	Sexual Arousal	Anxiety	Joy	Pain	Guilt
Nose	↓	↓		↑		↓		↓
Cheeks			↓					
Periorbital			↑	↑	↑			
Supraorbital			↑		↑			
Forehead	↓ ↑	↓		↑	↑		↓	
Maxillary	↓	↓	↓				↓	↓
Neck-carotid			↑					
Finger/palm							↓	
Lips/mouth				↑				

Similar results were found in a recent study by Cruz-Albarran et al., where thermal IR imaging was used during the emotions induction process to quantify temperature changes that occurred on different ROIs of the face [56]. The authors were able to classify the emotions relying on regional temperatures with an accuracy of 89.9%. The induced emotions were joy, disgust, fear, anger, and sadness. The examined ROIs were nose, cheeks, forehead, and maxillary area, which are depicted in Figure 2. Among all the ROIs, the nose and maxillary area were the most responsive to emotional stimuli, as they showed a significant change in temperature in all the induced emotions. The forehead temperature changed during sadness, anger, and fear, while the temperature of the cheeks changed during disgust and sadness.

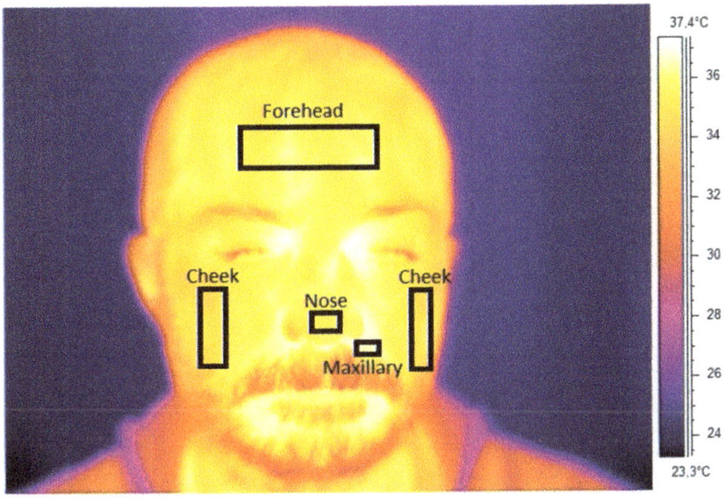

Figure 2. Thermal IR images with marked ROIs (black rectangles).

Emotion identification through thermal IR imaging was also employed in studies on children. Such applications may receive the greatest benefit from the technology thanks to the ecological nature of this technique and the difficulties associated with measuring children with skin-located sensing. Goulart et al. proposed an experimental design to identify five emotional states (disgust, fear, happiness, sadness, and surprise) evoked in children between 7 and 11 years old [57]. The forehead, the tip of nose, the cheeks, the chin, the periorbital and the perinasal regions were chosen to extract the affective information. Each thermal frame was processed by segmenting the ROIs and evaluating the ROIs' temperature mean, variance, and median values. Then, a linear discriminant analysis was used as a classifier. High accuracy (higher than 85%) was obtained for the classification of the five emotions, thus resulting in a robust method to identify quantitative patterns for emotion recognition in children. Temperature decrease was detected in the forehead region during disgust and surprise; in the periorbital region during happiness, sadness, and surprise; in the perinasal region during disgust and happiness; in the chin during surprise, happiness, and sadness; and finally, in the nose during disgust, fear, and happiness. The temperature increase was detected in the left cheek region for all emotions and in the nose tip during surprise.

Beyond the basic emotions, thermal IR imaging has been used to characterize the two dimensions of emotions, such as valence (pleasant versus unpleasant) and arousal (low versus high). Emotions dimensions are a crucial aspect in the affective research field, on which most of the studies on emotions recognition are based. The most commonly used models representing emotions dimension in HRI are the Pleasure, Arousal, Dominance (PAD) emotional state model [58] and the circumplex model of affect [59]. The PA dimensions of PAD were developed into the circumplex model, which indeed assume that any emotion might be described with two continuous dimensions of valence and arousal [60]. The valence dimension indicates whether the subject's current emotional state is positive or negative. Arousal, on the other hand, indicates whether the subject is responsive or not, at that given moment and for that given stimulus, and how active he/she is. In particular, the theory of the dimension of the emotions proposes that the emotional states are not discrete categories but rather a result of varying degrees of their dimensions. A graphical representation of the circumplex model of affect developed by Russel is reported in Figure 3. For example, joy is characterized as the product of strong activation in the neural systems associated with positive valence or pleasure together with moderate activation in the neural systems associated with arousal (i.e., low arousal). Emotions other than joy likewise arise from the same two-dimensional systems but differ in the degree or extent of activation. This allows characterizing also complex emotions such as love or happiness other than the basic ones. The analysis of the emotion recognition solutions reveals that there is no one commonly accepted standard model for emotion representation. The dimensional adaptation of Ekman's six basic emotions and the circumplex or PAD model are the ones widely adopted in emotion recognition solutions [61].

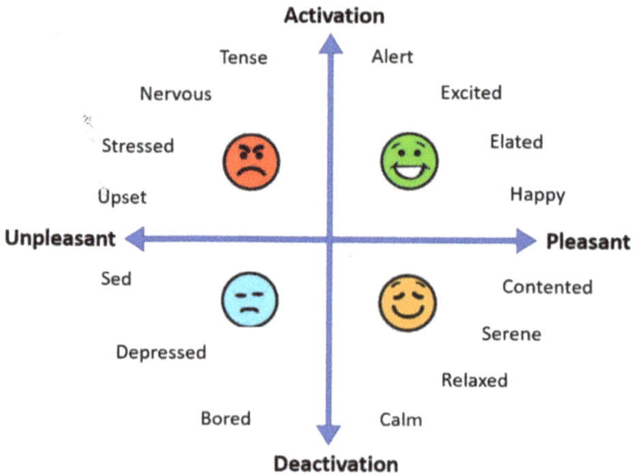

Figure 3. Graphical representation of the circumplex model of affect with the horizontal axis representing the valence dimension and the vertical axis representing the arousal dimension. Adapted from [60].

Recent studies explored temperature changes associated with different degrees of valence and arousal. For instance, Salazar-Lopez et al. studied the relation between changes in temperature of the subject's face and valence or arousal dimensions [53]. They used pictures from the International Affective Picture System (IAPS), which is widely used in studies of emotion recognition and characterized along the two dimensions [62]. The analyzed ROIs were the forehead (left and right sides), the tip of the nose, the cheeks, the eyes, and the mouth regions. Significant differences in temperature were found only on the tip of the nose. The results showed that high arousal images elicited temperature increases on the tip of the nose, while low arousal images led to temperature increases for pleasant images (i.e., positive valence) and decreases for unpleasant ones (i.e., negative valence). Contrasting results were indeed found by Kosonogov et al. [63]. The authors found no significant temperature differences along the valence dimension of the emotions (i.e., pleasant and unpleasant emotions). Besides, an activation effect of emotional pictures was found on the amplitude and latency of nasal thermal responses: the more arousing the pictures, the faster and larger the thermal responses. The only evaluated region was the tip of the nose. The relevance of the emotional arousal in causing changes in the nose tip temperature is supported and demonstrated in different studies, such as in Diaz-Piedra et al., where a direct relationship was found between reduced levels of arousal and nasal skin temperature [64], and confirmed in Bando et al. [65].

Moreover, Pavlidis et al. proposed quantifying stress through thermal IR imaging [54]. Stress in the valence–arousal space is identified as negative valence and high arousal [66]. Pavlidis et al. found that high stressful situations resulting in the cooling of the area around the nose tip [54]. Parallel results were found in children, Ioannou et al. obtained, in their study of guilt in children (age 39–42 months), in which the higher the distress signs, the higher the decrease in nose temperature [67]. The sense of guilt was induced through the "mishap paradigm" in which children were led to believe they had broken the experimenter's favorite toy (Figure 4). The temperature of the nasal area decreased following the "mishap" condition, suggesting a sympathetic activation and peripheral nasal vasoconstriction, and it increased after soothing due to a parasympathetic activation.

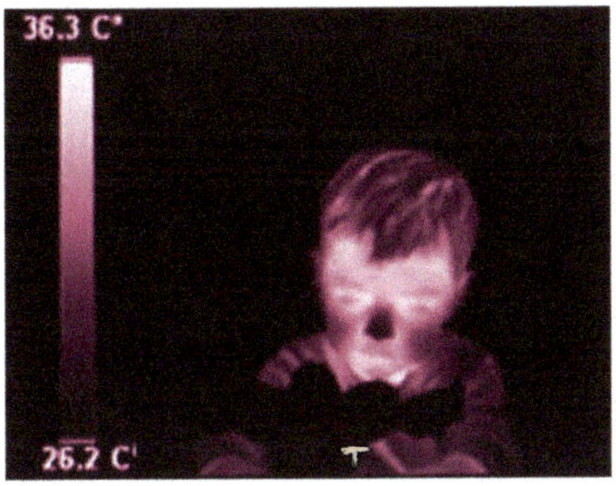

Figure 4. Child facial temperature during the mishap condition. Adapted from [68].

Vasomotor process, subcutaneous blood flow, and perspiration patterns are not the only physiological measures that can be detected through thermal IR imaging. In fact, the physiological signature of the autonomic system activation such as heart and breathing rate variation can be also monitored [69–72]. These two parameters are measurable with thermal IR imaging by positioning a ROI over a superficial vessel or over the nostrils respectively and by monitoring the ROI average temperature over time. Whereas the latter is easily detectable, because the thermal difference between the expired and inspired air is easily appreciable (approximately 0.5 °C), the modulation of temperature caused by the pulsation of blood in the vessel is not. Therefore, a series of algorithms have been developed for the estimation of the heart rate through thermal imaging [69]. Cross et al. used thermal IR imaging to detect physiological indicators of stress in the adult population by analyzing the respiration and heart rate variation during the performance of mental and physical tasks [26]. Temperature variation over time was recorded on the nose tip and regions near superficial arteries to detect respiration and heart rates, respectively. The results showed that the accuracy in the physical versus psychological stressors classification was greater than 90%, and the heart and respiration rate were accurately detected by thermal IR imaging. Whereas the evaluation of the heart rate through thermal IR imaging is not common in the literature, the respiration rate assessment through thermal IR imaging has been tested in different settings, including sleep monitoring [73], neonatal care [74], and driver's drowsiness monitoring [75].

5. Limits of Current Thermal IR Imaging for HRI Applications

Studies reviewed in this section revealed thermal IR imaging ability to monitor physiological signs and affective states. Although they have sometimes shown incongruent results (e.g., nose tip temperature did not significantly change along the valence dimension of the emotions in Kosonogov et al. [63], whilst it did change in Salazar-López et al. [53]), these findings open exciting prospects for affective computing. One of the causes of inconsistency could be that a single discrete metric is maybe not always sufficient, since it could be susceptible to the complex physiological mechanisms [76]. In addition, there could be interaction effects between affective states. Nonetheless, all the affective states analyzed have the capability to induce ROI temperature variations. However, an important consideration is that all these studies were conducted using high-end thermal camera and performed in controlled laboratory settings. The thermal systems mostly used in literature are FLIR (Wilsonville, OR, USA) A655sc with 640 × 480 spatial resolution, a 50 Hz sampling rate, and <0.03 °C thermal sensitivity and a FLIR A325sc with 320 × 240 spatial resolution, 60 Hz sampling rate, and <0.05 °C thermal sensitivity. In addition, all the observations reported were made in a climate-controlled room according to the International Academy of Thermology (IACT) guidelines [77]. In fact, IACT guidelines indicated that when performing a thermal IR imaging measurement, it is mandatory to control the temperature and humidity of the experimental room. They suggested a temperature range of 18–23 °C and a controlled humidity range between 40% and 70%. No direct ventilation on the subject and no direct sunlight (no windows or with curtains or blinds) is also recommended during the experimental measures. In conclusion, by analyzing the studies reported in this section, two main constraints were identified that are not suitable for HRI applications. Those are (1) the use of high-end and sized thermal imaging systems and (2) the circumstances in which the studies were conducted i.e., in restricted laboratory settings. On the other hand, HRI applications require daily life scenarios, eventually suitable for outdoor use, and technology embedded in commercial social robots, i.e., low-cost miniaturized sensors. In Sections 6 and 7, those limits are addressed in order to highlight the new improvements developed in recent studies. A special emphasis was placed on these two sections as they deal with crucial aspects for the use of thermal IR imaging in the HRI field. Finally, the last section focuses on the actual state of the art of thermal IR imaging-based affective computing applications.

6. Mobile Thermal IR Imaging

The relevant spread of thermal IR technology, together with the miniaturization of IR detectors, induced manufacturers to produce portable thermographic systems, i.e., mobile and low-cost thermal IR imaging devices. One of the first companies to commercialize mobile thermal devices was FLIR with systems such as the FLIR ONE Pro, 160 × 120 spatial resolution, an approximately 8.7 Hz sampling rate, 0.15 °C thermal sensitivity, and dimensions of 68 × 34 × 14 mm^3, or FLIR Lepton, 80 × 160 spatial resolution, an approximately 8.7 Hz sampling rate, <0.05 °C thermal sensitivity, and dimensions of 11.8 × 12.7 × 7.2 mm^3. FLIR ONE was designed to be integrated on mobile phones. Another mobile thermal system designed to be integrated on a mobile phone is the Therm-App system developed by Opgal manufacturer; it has 384 × 288 spatial resolution, an approximately 9 Hz sampling rate, approximately 0.07 °C thermal sensitivity, and dimensions of 55 × 65 × 40 mm^3. Market research has shown that the SmartIR640 mobile thermal system, manufactured by Device aLab (640 × 480 spatial resolution, 30 Hz sampling rate, <0.05 °C thermal sensitivity, and dimensions of 27 × 27 × 18 mm^3) is also a valid solution, but it is not yet used for research projects.

Despite the relatively low-quality thermal imaging outputs of a mobile thermal system, this technology could help bridge the gap between the findings from highly constrained laboratory environments and wild real-world applications. Indeed, its portability (e.g., small size and low computational resource requirement) allows the camera not only to be easily attached to a mobile phone but also to be integrated in a social robot head. Recent studies have started to explore mobile thermal IR imaging for affect recognition tasks, especially focused on stress monitoring [78–81]. Cho et al. proposed a system consisting of a smartphone camera-based PhotoPlethysmoGraphy (PPG) and a low-cost thermal camera added to the smartphone, which was designed to continuously monitor the subject's mental stress [78]. By analyzing the nose tip temperature and the blood volume pulse through PPG [82,83], they were able to classify the stress with an accuracy of 78.33%, which is comparable to the state-of-the-art stress recognition methods. The employed mobile thermal camera was the FLIR ONE. The study was conducted in a quiet laboratory room with no distractions. Another study by the same authors included the mobile thermal camera as a standalone system to monitor mental stress [80]. The authors proposed a novel low-cost non-contact thermal IR imaging-based stress recognition system that relayed on the breathing dynamic patterns analysis. In fact, since breathing is an important vital process controlled by the ANS, its pattern monitoring can be informative of a person's mental and physical condition. Results showed a classification accuracy of 84.59% for a binary classification (i.e., no-stress, stress) and an accuracy of 56.52% for multi-class classification (i.e., none, low, high-level stress). The breathing signal was recovered by tracking the person's nostril area. Then, the extracted breathing signal was converted in a two-dimensional spectrogram by stacking the Power Spectral Density (PSD) vector of a short-time-window respiration signal over time. Since the PSD function handles the short-time autocorrelation that identifies similarities between neighboring signal patterns, it can be used to examine respiration variations in a short period [65]. The breathing signal was also investigated through the use of a mobile thermal camera by Ruminski et al., who embedded such a camera in smart glasses [84]. Basu et al. instead used a mobile thermal system for the challenging purpose of classifying personality (psychoticism, extraversion, and neuroticism) [81]. The proposed system classified the emotional state using an information fusion of thermal and visible images. A blood flow perfusion model was used to obtain discriminating eigenfeatures from the thermal system. Then, these eigenfeatures were fused with those of visible images and classified. The blood perfusion model was obtained by analyzing the thermogram of the entire face and using Pennes' bioheat equation. The classification performance reached 87.87%.

Summarizing, mobile thermal IR imaging can provide high levels of flexibility and suitability for recovering physiological signatures such as the breathing signal and recognizing a person's affective states. However, a key challenge on the use of a mobile thermal system is related to the low quality of the output signal due to the low thermal and spatial resolution of the imaging system. The low spatial resolution can be easily addressed by bringing the thermal camera closer to the region of interest, but this is not always possible. An interesting method to overcome such an issue is presented in Cho et al. [85]. The authors proposed the *Thermal Gradient Flow* and *Thermal Voxel Integration* algorithms. *Thermal Gradient Flow* was mainly based on building thermal-gradient magnitude maps for enhancing the boundary around the region of interest, which in turn contributes to making the system robust to motion artifacts in presence of low-resolution images. Instead, the *Thermal Voxel Integration* consisted of a projection of a 2D thermal matrix onto a 3D space by taking a unit thermal element as a thermal voxel. This method was applied for breathing signal analysis application, and it resulted in producing a higher quality of breathing patterns.

7. Thermal IR Imaging-Based Affective Computing Outside Laboratory Settings

The market entry of smaller and low-cost thermal cameras is paving the way for thermal IR imaging applications outside laboratory environments. However, only few studies have been conducted. One example is reported by Goulart et al., who proposed a camera system composed of thermal and visible cameras for emotion recognition in children [86]. The camera system was attached to the head of a social robot, and the experiment was conducted in a room within the children's school environment. The room temperature was kept between 20 and 24 °C, using a constant luminous intensity. To capture thermal variation in the children's faces, they used the mobile thermal camera Therm-App. A similar set-up has been reported in Filippini et al. [87]. The authors installed a mobile thermal IR imaging system, the FLIR ONE, on the head of a social robot. The study was conducted in a primary school. Filippini and Goulart's studies are described in detail in the next section. Although both experiments were performed outside a laboratory setting, they still had constraints that are not adaptable to all real-life applications, such as the need to maintain a stable temperature, which is not possible in open air contexts or applications. Instead, Cho et al. conducted one of their experiments in unconstrained settings with varying thermal dynamic range scenes (i.e., indoor and outdoor physical activity). The experiment was aimed to monitoring a person's thermal signatures while walking [85].

Concerning the employment of thermal mobile system for affective computing purposes, one of the most compelling challenges for real-world environment is to ensure reliable thermal tracking of the chosen ROIs. This is due to dynamic changes in ambient temperature that affect the skin and can cause an inconsistent thermal signal coupled with the low resolution of the low-cost thermal system. This aspect makes it difficult to track ROIs automatically. Moreover, applications in real-world scenarios require real-time responses from the sensors of interest. Hence, an automatic recording of thermal IR imaging data and real-time processing is required. In this regard, signal processing techniques need to be chosen based on their efficiency in terms of computational load to allow acceptable performance for real-time processing. In Goulart et al. and Filippini et al., the tracking algorithm in thermal images relied on the visible images [86,87]. Thereby, the first phase consisted on the camera calibration done through a synchronous acquisition between visual and thermal images and using a checkerboard whose details were clearly detectable by both the visible spectrum and the thermal camera. After the calibration, the face detection and ROIs localization were performed on the visible image, since a state-of-the-art computer vision algorithm can be used in the visible field, and then the ROI coordinates in the visible images were converted to those of the thermal image. In Goulart et al., the Viola–Jones algorithm was used for visible ROIs detection, and then these ROIs were transferred to the corresponding thermal camera frame through a homography matrix [86]. In Filippini et al., the authors used an object detector based on the histogram of oriented gradients (HOG) for face localization in visible images [87]. The extraction of landmarks was based, instead, on a regression tree ensemble algorithm. An average of 82.75% of the total amount of the frames were correctly tracked as a result. However, it is important to mention that a further improvement would be to develop a real-time tracker, based on the only IR videos, acquired by low-resolution thermal cameras, to avoid problems due to a low-light environment and to infer the psychophysiology state of the human interlocutor. An attempt in this direction was proposed in Cho et al., where the authors were interested in tracking the physiological signal related to the only breathing process, focusing on the nostril region [85]. The approach they proposed was intended to compensate for the effects of variation in ambient temperature and movement artifacts, and it was named the "Optimal Quantization" method. The quantization process itself is the process of translating from a continuous temperature value to its digital color-mapped equivalent. This method consisted in adaptively quantizing the thermal distribution sequences by finding a thermal range of interest that contains the whole facial temperature distributions for every single frame. Despite the enhancing approach, this cannot cover all possible scenarios such as contexts of high humidity or severe temperature condition. Nonetheless, the further development of automatic ROI tracking on thermal images in entirely mobile and ubiquitous situations is required.

8. Thermal IR Imaging-Based Affective Computing in HRI

For decades, a growing interest on the possibility of developing intelligent machines that engage in social interaction has been observed. Many researchers, also in the field of thermal IR imaging, have been enthralled with the HRI appeal and with the possibility of developing robots that are capable of social interactions with humans. The application of thermal IR imaging in this field was firstly investigated by Merla in 2014 [88]. Since then, many studies have been conducted to validate the thermal IR imaging technique for emotion recognition analysis with the aim of applying this technique in the HRI field. However, only few applications have been actually implemented in this area. One of the first attempts was carried out by Sorostinean et al. [89]. In this study, the authors presented the design of a thermal IR imaging-based system mounted on a humanoid robot performing a contact-free measurement of temperature variations across people's faces in a stressful interaction. The results showed a statistically significant interaction between the distance and the gaze direction of the robot and the temperature variation of the nasal and peri-nasal region. This supported the fact that thermal imaging sensors can be successfully employed in embodying robots with physiological sensing capabilities to allow them to become aware of their effect on people, know about their preferences, and build a reactive behavior. Agrigoroaie et al. reported promising preliminary results in their attempt to determine if a person was trying to deceive a robot through the use of a mobile thermal camera [90]. Instead, Boccanfuso et al. evaluated the efficacy of the thermal IR imaging for detecting robot-elicited affective response compared to video-elicited affective response by tracking thermal changes in five ROIs on the subject's face [91]. They studied the interaction effects of condition (robot/video) and emotion (happy/angry) on individual facial ROIs. Although no interaction effects for most ROI temperature slopes were found, a strong, statistically significant effect of the interaction between condition and emotion when evaluating the temperature slope on the nose tip was observed. This result confirms again the assumption that the nasal area is a salient region for emotion detection [32,67,92–95]. Recent studies included applications with more challenging populations such as infants and children. Scassellati et al. proposed the design of a unique dual-agent system that uses a virtual human and a physical robot to engage 6–12-month-old deaf infants in linguistic interactions. The system was endowed with a perception system that was capable of estimating infant attention and engagement through thermal imaging and eye tracking [96]. This study was part of a larger project designed to develop a system called RAVE (Robot AVatar thermal Enhanced language learning tool). RAVE is aimed to be an augmentative learning tool that can provide linguistic input, in particular visual language inputs, to facilitate language learning during one widely recognized critical developmental period for language (ages 6–12 months [97]) [96,98–101]. To this end, thermal IR imaging was used to determine the emotional arousal and attentional valence, providing new knowledge about when infants are most optimally "Ready to Learn", even before the onset of language production. This is prominent for infants who might not otherwise receive sufficient language exposure. Of particular concern are deaf babies, many of whom are born to parents who do not know a signed language [96].

Although these studies were very ambitious and fascinating, they were still carried out in constrained laboratory settings, probably implying a not entirely free interaction. To the state of the art, the only two studies performed so far that have concerned HRI applications in a out-of-laboratory context are those reported in Section 7 [86,87].

Goulart et al. used a mobile social robot called N-MARIA (New-Mobile Autonomous Robot for Interaction with Autistics), which was built in UFES/Brazil to assist children during social relationship rehabilitation [86]. In the interaction with the child, which lasted two minutes, the child was encouraged to make communication and tactile interaction with the robot. The robot was equipped with low-cost hardware (thermal and visible camera) used to give information about the emotional state of the child, which was related to five emotions (i.e., surprise, fear, disgust, happiness, and sadness). Results showed that the system was able to recognize those emotions, achieving 85.75% accuracy. Such accuracy was comparable with gold standard techniques in emotion recognition, such as facial expression analysis and speech tone analysis [102–106]. Filippini et al. used the social robot "Mio Amico" robot, produced by ©Liscianigiochi, in which the mobile thermal system FLIR ONE (which includes a thermal and a visible camera) was installed on the head of the robot [87]. The study aimed to endow the robot with the capability of real-time assessment of the interlocutor's state of engagement (positive, neutral, and negative emotional engagement). During the interaction between the robot and the child, the robot could either tell a fairy tale or sing a song. At the end of the fairy tale or the song, the robot asked the child if he/she liked the fairy tale/song and if he wanted to listen to another one. Based on the child's answer, the robot could choose the next action. The engagement state of the infant was classified by analyzing the child's thermal modulation using a low computational processing pipeline. Figure 5 summarizes in a–c the processing pipeline for the interlocutor's state of engagement identification and (d) the child–robot interaction. The accuracy reached was 70%. Although this study presented a lower level of accuracy compared to Goulart et al. [86], it is worth mentioning that the estimation of the level of engagement can be considered a hard task, since it represents a complex emotion (combination of the basic ones), and it has been poorly investigated in the literature. Besides, the study reported in Filippini et al. [87] represents the first and unique study in which the robot could actually change its activities based on the child affective state, opening the way to a bidirectional interaction.

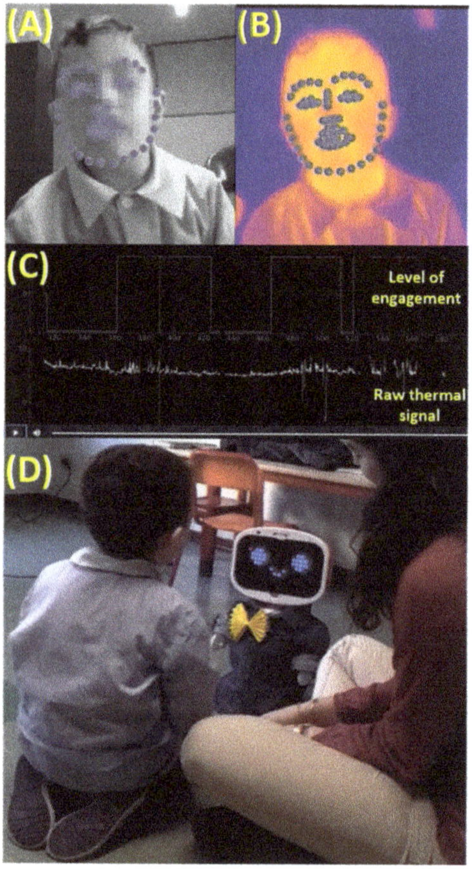

Figure 5. (**A–C**) Processing pipeline. (**A**) The first phase relied on the visible image to detect the child's face and locate specific facial landmarks. (**B**) The corresponding landmarks on the thermal image were then obtained thanks to a previous optical calibration procedure. (**C**) The thermal signal is extracted from the nose tip region and processed in order to obtain the subject's level of engagement. (**D**) The child–robot interactions are guided by the robot's understanding of the child's level of engagement.

9. Thermal IR Imaging-Based Affective Computing in Intelligent Systems Such as Driver-Assistance Systems or Autonomous Vehicles

Besides humanoid robots, systems such as driver-assistance systems or autonomous vehicles can also be classified as "robots" due to their intelligent behaviors. Indeed, driving functions are becoming increasingly automated; consequently, motorists could run the risk of being cognitively removed from the driving process. Thermal IR imaging was demonstrated to be a valid technique for assessing variation in cognitive load [52,107,108]. Most of all, it could enable sensing the real-time state of motorists non-invasively i.e., without disrupting driving-related tasks and, unlike RGB cameras, independently from external light conditions [109]. Studies of thermal IR imaging in driving contexts stated that a rise in mental workload leads to an increase in the difference between nose and forehead temperature [108,110]. Moreover, thermal IR imaging was demonstrated to be a valuable indicator of the driver arousal level, from alertness to drowsiness [64,111]. A recent study employed it to detect human thermal discomfort in order to develop a fully automated climate control system in the vehicles [112]. Even in the view of automated vehicles, in complex situations, they will need human assistance for a long time. Therefore, for road safety, driver monitoring is more relevant than ever, in order to keep the driver alert and awake.

Autonomous driving relies on the understanding of objects and scenes through images. A recent study assessed a fusion system consisting of a visible and thermal sensor on object recognition from a driving dataset, and it demonstrated that thermal images significantly improve detection accuracy [113]. Miethig et al. argued that thermal IR imaging can improve the overall performance of existing autonomous detection system and reduce pedestrian detection time [114].

In conclusion, thermal IR imaging can be greatly useful also in vehicle technology; however, further testing is needed to better understand how it would improve automated vehicles and the knowledge about cognitive states in traffic safety.

10. Discussion

The advancement in robotics, especially in social robotics, is breathtaking. Social robots could potentially revolutionize the way of taking care of the sick or the elderly people, the way of teaching and learning, and even the definition of the concept of companionship. Nowadays, social robots hold the promise of extending life expectancies and improving health and the quality of life. In this review, the impact that social robots have in fields such as education, healthcare, and tourism is briefly surveyed. In all these areas, they are mainly intended to improve or protect the lifestyle and health, both physical and mental, of the human user. In fact, robots can also help socially impaired people relate to others, practice empathetic behaviors, and act as a steppingstone toward human contact. However, the most important and desirable requirement is that the robot meets the person's needs. Robots actually need to be able to recognize the interlocutor's affective state to communicate naturally with him/her and to engage him/her not only on a cognitive level but on an emotional level as well.

To be able to get information about the partner's affective state, there are at least two possibilities. Either the user has to explicit and voluntarily express information about his/her emotions to the robot—for example, using natural language or facial expressions and gestures, or the robot has to recognize involuntary affective information from physiological measures, such as respiration, heart rate, skin conductance, skin temperature, and blood pressure. In the present review, the use of an ecological technology is promoted, such as the thermal IR imaging-based affective computing technique. It aimed to facilitate HRI by endowing the robots with the capability of autonomously identifying the person's emotional state. Thermal IR imaging, by recording facial cutaneous temperature and its topographic distribution, is able to identify specific features clearly correlated to emotional state and measures associated with standard physiological signals of the sympathetic activity. Emotional state, in fact, determines a redistribution of the blood in the vessels through vasodilation or vasoconstriction phenomena, which are regulated by ANS. These phenomena can be identified and monitored over time because they produce a thermal variation on the skin. The thermal IR imaging technique is already validated in the literature for emotion recognition tasks.

Regarding emotion recognition, gold standard techniques such as speech and facial expression analysis were mentioned. However, it is worth mentioning that to distinguish between emotions, different models or theories have been so far developed and used by psychologists or cognitive neuroscientists. Thus, it is difficult to take a theory of one research field, such as psychology, and apply it to another, such as HRI. This remains an open discussion issue in the HRI research field. For instance, in speech analysis, the emotion recognition models developed using the utterances of a particular language usually do not yield appreciably good recognition performance for utterances from other languages [115]. On the other hand, in facial expression and gesture analysis, two main theories are currently established in emotion research: a discrete approach, claiming the existence of universal "basic emotions" [116], and a dimensional approach, assuming the existence of two or more major dimensions that can describe different emotions and distinguish between them [117]. The thermal IR imaging technique, i.e., the ROI's temperature modulation analysis, has demonstrated to be suitable for emotion recognition based on both the basic emotion approach and on the dimensional theory of the emotion (as reported in Section 4). This makes thermal IR imaging a cross-cutting and ubiquitous technique in the area of emotion recognition and consequently a valuable contribution in HRI studies. Of course, thermal IR imaging is not the first and unique technology explored for emotion recognition through physiological measures in HRI, but it seems to be one of the most ecological. In this review, the major challenges toward the application of this technique in HRI fields has been highlighted, bridging the gap between the constrained laboratory setting and the real-world scenario. To this end, a few studies have been reported in the literature and were here analyzed. The application of thermal IR imaging-based Affective Computing in the field of HRI has been reviewed as well. Interesting results were reported by Goulart et al, in which the developed system was able to recognize emotions, achieving 85.75% accuracy [86]. To some extent, such accuracy can be comparable with gold standard techniques in emotion recognition for HRI, such as facial expression analysis and speech tone analysis.

An important aspect that further draws attention to this technique is its adaptability in applications where there is interaction between social robots and challenging populations such as neonates. This was the case of RAVE, the learning tool composed of an avatar and a social robot, which was designed to facilitate langue learning during the critical developmental period (age 6–12 months) and devolved to infants who might not otherwise receive proper language exposure. In conclusion, we believe that this review paves the way for the use of thermal IR imaging in HRI, which could endow the social robot with the capability of recognizing the interlocutor's emotions relying on involuntary physiological signals measurements. These measurements may be fed to multivariate linear [118,119] or non-linear regressors or classification algorithms [120], also relying on data-driven machine learning and deep learning approaches [121,122]. In this way, it is possible to avoid the artifact of social masking and make HRI suitable also for people who lack the ability to express emotions. Beyond robots, intelligent systems such as autonomous vehicles or even smart buildings could also benefit from this technique.

11. Conclusions

HRI is a relatively young discipline that has attracted a lot of attention in recent years due to the increasing availability of complex robots and people's exposure to such robots in their daily lives. Moreover, robots are increasingly being developed for real-world application areas, such as education, healthcare, eldercare, and other assistive applications. A natural HRI is crucial for the beneficial influence that robots can have in human life. Understanding the interlocutor's need and affective state during the interplay is the foundation of a human-like interaction. To this end, an ecological technology such as thermal IR imaging, which can provide information about physiological parameters associated to the subject affective state, was here presented and surveyed. The technology can provide the ground for the further development of robust social robots and to facilitate HRI. Thermal IR imaging has already been validated in the literature in the fields of emotion recognition. This review can act as a guideline to, and foster, the use of thermal IR imaging-based affective computing in HRI applications, which is intended to support a natural HRI, with special regard to those who find difficult to express emotions.

Author Contributions: Conceptualization, C.F., A.M.; investigation, C.F., D.C., A.M.C., D.P.; writing—original draft preparation, C.F.; writing—review and editing D.C., A.M.C., D.P.; supervision, A.M.; funding acquisition, A.M. All authors have read and agreed to the published version of the manuscript.

Funding: This research was funded by PON FESR MIUR R&I 2014-2020 - Asse II - ADAS+, ARS01_00459 and PON MIUR SI-ROBOTICS, ARS01_01120.

Conflicts of Interest: The authors declare no conflict of interest.

References

1. Sung, J.-Y.; Guo, L.; Grinter, R.E.; Christensen, H.I. "My Roomba is Rambo": Intimate home appliances. In Proceedings of the International Conference on Ubiquitous Computing, Innsbruck, Austria, 16–19 September 2007; Springer: Berlin/Heidelberg, Germany, 2007; pp. 145–162.
2. Dautenhahn, K. Methodology & themes of human-robot interaction: A growing research field. *Int. J. Adv. Robot. Syst.* **2007**, *4*, 15.
3. Salter, T.; Werry, I.; Michaud, F. Going into the wild in child–robot interaction studies: Issues in social robotic development. *Intell. Serv. Robot.* **2008**, *1*, 93–108. [CrossRef]
4. Horvitz, E.; Paek, T. Harnessing models of users' goals to mediate clarification dialog in spoken language systems. In Proceedings of the International Conference on User Modeling, Sonthofen, Germany, 13–17 July 2001; Springer: Berlin/Heidelberg, Germany, 2001; pp. 3–13.
5. Kahn, P.H., Jr.; Ishiguro, H.; Friedman, B.; Kanda, T.; Freier, N.G.; Severson, R.L.; Miller, J. What is a Human? Toward psychological benchmarks in the field of human–robot interaction. *Interact. Stud.* **2007**, *8*, 363–390.
6. Duffy, B.R. Anthropomorphism and the social robot. *Robot. Auton. Syst.* **2003**, *42*, 177–190. [CrossRef]
7. Turkle, S. In good company? On the threshold of robotic companions. In *Close Engagements with Artificial Companions*; John Benjamins: Amsterdam, The Netherlands, 2010; pp. 3–10.
8. Thomasen, K. Examining the constitutionality of robot-enhanced interrogation. In *Robot Law*; Edward Elgar Publishing: Cheltenham, UK, 2016.
9. Darling, K. Extending legal protection to social robots: The effects of anthropomorphism, empathy, and violent behavior towards robotic objects. In *Robot Law*; Edward Elgar Publishing: Cheltenham, UK, 2016.
10. Darling, K. 'Who's Johnny?' Anthropomorphic Framing in Human-Robot Interaction, Integration, and Policy. *Anthr. Fram. Hum. Robot Interact. Integr. Policy March 23 2015 ROBOT ETHICS* **2015**, *2*. [CrossRef]
11. Press, T.M. Affective Computing. Available online: https://mitpress.mit.edu/books/affective-computing (accessed on 27 December 2019).
12. Russell, J.A.; Bachorowski, J.-A.; Fernández-Dols, J.-M. Facial and vocal expressions of emotion. *Annu. Rev. Psychol.* **2003**, *54*, 329–349. [CrossRef] [PubMed]
13. Fernandes, A.; Helawar, R.; Lokesh, R.; Tari, T.; Shahapurkar, A.V. Determination of stress using blood pressure and galvanic skin response. In Proceedings of the 2014 International Conference on Communication and Network Technologies, 2014, Sivakasi, India, 18–19 December 2014; pp. 165–168.
14. Bradley, M.M.; Lang, P.J. Measuring emotion: Behavior, feeling, and physiology. *Cogn. Neurosci. Emot.* **2000**, *25*, 49–59.
15. Schachter, S.; Singer, J. Cognitive, social, and physiological determinants of emotional state. *Psychol. Rev.* **1962**, *69*, 379. [CrossRef]
16. Knapp, R.B.; Kim, J.; André, E. Physiological signals and their use in augmenting emotion recognition for human–machine interaction. In *Emotion-Oriented Systems*; Springer: Berlin/Heidelberg, Germany, 2011; pp. 133–159.
17. Reynolds, C.; Picard, R. Affective sensors, privacy, and ethical contracts. In Proceedings of the CHI'04 Extended Abstracts on Human Factors in Computing Systems, 2004, Vienna, Austria, 24–29 April 2004; pp. 1103–1106.
18. Sebe, N.; Sun, Y.; Bakker, E.; Lew, M.S.; Cohen, I.; Huang, T.S. Towards authentic emotion recognition. In Proceedings of the 2004 IEEE International Conference on Systems, Man and Cybernetics (IEEE Cat. No. 04CH37583), Hague, The Netherlands, 10–13 October 2004; Volume 1, pp. 623–628.
19. Schmidt, K.L.; Cohn, J.F. Human facial expressions as adaptations: Evolutionary questions in facial expression research. *Am. J. Phys. Anthropol. Off. Publ. Am. Assoc. Phys. Anthropol.* **2001**, *116*, 3–24. [CrossRef]
20. Crivelli, C.; Fridlund, A.J. Facial displays are tools for social influence. *Trends Cogn. Sci.* **2018**, *22*, 388–399. [CrossRef]
21. Wioleta, S. Using physiological signals for emotion recognition. In Proceedings of the 2013 6th International Conference on Human System Interactions (HSI), 2013, Sopot, Poland, 6–8 June 2013; pp. 556–561.
22. Jerritta, S.; Murugappan, M.; Nagarajan, R.; Wan, K. Physiological signals based human emotion recognition: A review. In Proceedings of the 2011 IEEE 7th International Colloquium on Signal Processing and its Applications, 2011, Penang, Malaysia, 4–6 March 2011; pp. 410–415.

23. Basu, A.; Routray, A.; Shit, S.; Deb, A.K. Human emotion recognition from facial thermal image based on fused statistical feature and multi-class SVM. In Proceedings of the 2015 Annual IEEE India Conference (INDICON), 2015, New Delhi, India, 17–20 December 2015; pp. 1–5.
24. Puri, C.; Olson, L.; Pavlidis, I.; Levine, J.; Starren, J. StressCam: Non-contact measurement of users' emotional states through thermal imaging. In Proceedings of the CHI'05 extended abstracts on Human factors in computing systems, 2005, Portland, OR, USA, 2–7 April 2005; pp. 1725–1728.
25. Wang, S.; He, M.; Gao, Z.; He, S.; Ji, Q. Emotion recognition from thermal infrared images using deep Boltzmann machine. *Front. Comput. Sci.* **2014**, *8*, 609–618. [CrossRef]
26. Cross, C.B.; Skipper, J.A.; Petkie, D.T. Thermal imaging to detect physiological indicators of stress in humans. In Proceedings of the Thermosense: Thermal Infrared Applications XXXV, Baltimore, MD, USA, 30 April–2 May 2013; International Society for Optics and Photonics: Bellingham, WA, USA, 2013; Volume 8705, p. 87050.
27. Cardone, D.; Pinti, P.; Merla, A. Thermal infrared imaging-based computational psychophysiology for psychometrics. *Comput. Math. Methods Med.* **2015**, *2015*. [CrossRef] [PubMed]
28. Engert, V.; Merla, A.; Grant, J.A.; Cardone, D.; Tusche, A.; Singer, T. Exploring the Use of Thermal Infrared Imaging in Human Stress Research. *PLoS ONE* **2014**, *9*. [CrossRef] [PubMed]
29. Ebisch, S.J.; Aureli, T.; Bafunno, D.; Cardone, D.; Romani, G.L.; Merla, A. Mother and child in synchrony: Thermal facial imprints of autonomic contagion. *Biol. Psychol.* **2012**, *89*, 123–129. [CrossRef]
30. Paolini, D.; Alparone, F.R.; Cardone, D.; van Beest, I.; Merla, A. "The face of ostracism": The impact of the social categorization on the thermal facial responses of the target and the observer. *Acta Psychol. (Amst.)* **2016**, *163*, 65–73. [CrossRef]
31. Di Giacinto, A.; Brunetti, M.; Sepede, G.; Ferretti, A.; Merla, A. Thermal signature of fear conditioning in mild post traumatic stress disorder. *Neuroscience* **2014**, *266*, 216–223. [CrossRef]
32. Panasiti, M.S.; Cardone, D.; Pavone, E.F.; Mancini, A.; Merla, A.; Aglioti, S.M. Thermal signatures of voluntary deception in ecological conditions. *Sci. Rep.* **2016**, *6*, 1–10. [CrossRef]
33. Aureli, T.; Grazia, A.; Cardone, D.; Merla, A. Behavioral and facial thermal variations in 3-to 4-month-old infants during the Still-Face Paradigm. *Front. Psychol.* **2015**, *6*. [CrossRef]
34. Perpetuini, D.; Cardone, D.; Filippini, C.; Chiarelli, A.M.; Merla, A. Modelling Impulse Response Function of Functional Infrared Imaging for General Linear Model Analysis of Autonomic Activity. *Sensors* **2019**, *19*, 849. [CrossRef]
35. Perpetuini, D.; Cardone, D.; Chiarelli, A.M.; Filippini, C.; Croce, P.; Zappasodi, F.; Rotunno, L.; Anzoletti, N.; Zito, M.; Merla, A. Autonomic impairment in Alzheimer's disease is revealed by complexity analysis of functional thermal imaging signals during cognitive tasks. *Physiol. Meas.* **2019**, *40*, 034002. [CrossRef]
36. Kitchenham, B. *Procedures for Performing Systematic Reviews*; Keele University: Keele, UK, 2004; Volume 33, pp. 1–26.
37. Belpaeme, T.; Kennedy, J.; Ramachandran, A.; Scassellati, B.; Tanaka, F. Social robots for education: A review. *Sci. Robot.* **2018**, *3*. [CrossRef]
38. Girotto, V.; Lozano, C.; Muldner, K.; Burleson, W.; Walker, E. Lessons Learned from In-School Use of rTAG: A Robo-Tangible Learning Environment. In Proceedings of the 2016 CHI Conference on Human Factors in Computing Systems, Association for Computing Machinery, New York USA. San Jose, CA, USA, 7–12 May 2016; pp. 919–930.
39. Movellan, J.; Eckhardt, M.; Virnes, M.; Rodriguez, A. Sociable robot improves toddler vocabulary skills. In Proceedings of the 4th ACM/IEEE International Conference on Human Robot Interaction, La Jolla, CA, USA, 9–13 March 2009; pp. 307–308.
40. Tanaka, F.; Isshiki, K.; Takahashi, F.; Uekusa, M.; Sei, R.; Hayashi, K. Pepper learns together with children: Development of an educational application. In Proceedings of the 2015 IEEE-RAS 15th International Conference on Humanoid Robots (Humanoids), Seoul, Korea, 3–5 November 2015; pp. 270–275.
41. Banks, M.R.; Willoughby, L.M.; Banks, W.A. Animal-assisted therapy and loneliness in nursing homes: Use of robotic versus living dogs. *J. Am. Med. Dir. Assoc.* **2008**, *9*, 173–177. [CrossRef] [PubMed]
42. Hodges, H.; Fealko, C.; Soares, N. Autism spectrum disorder: Definition, epidemiology, causes, and clinical evaluation. *Transl. Pediatr.* **2020**, *9*, S55. [CrossRef] [PubMed]
43. Olaronke, I.; Oluwaseun, O.; Rhoda, I. State of The Art: A Study of Human-Robot Interaction in Healthcare. *Int. J. Inf. Eng. Electron. Bus.* **2017**, *9*, 43–55. [CrossRef]

44. Cabibihan, J.-J.; Javed, H.; Ang, M.; Aljunied, S.M. Why robots? A survey on the roles and benefits of social robots in the therapy of children with autism. *Int. J. Soc. Robot.* **2013**, *5*, 593–618. [CrossRef]
45. Sartorato, F.; Przybylowski, L.; Sarko, D.K. Improving therapeutic outcomes in autism spectrum disorders: Enhancing social communication and sensory processing through the use of interactive robots. *J. Psychiatr. Res.* **2017**, *90*, 1–11. [CrossRef]
46. Wilson, K.P.; Landa, R.J. Barriers to Educator Implementation of a Classroom-Based Intervention for Preschoolers with Autism Spectrum Disorder. *Front. Educ.* **2019**, *4*. [CrossRef]
47. Tussyadiah, I.P.; Park, S. Consumer Evaluation of Hotel Service Robots. In Proceedings of the Information and Communication Technologies in Tourism, Jönköping, Sweden, 24–26 January 2018; Stangl, B., Pesonen, J., Eds.; Springer International Publishing: Cham, Switzerland, 2018; pp. 308–320.
48. Niculescu, A.I.; Jiang, R.; Kim, S.; Yeo, K.H.; D'Haro, L.F.; Niswar, A.; Banchs, R.E. SARA: Singapore's Automated Responsive Assistant, A Multimodal Dialogue System for Touristic Information. In Proceedings of the Mobile Web Information Systems, Barcelona, Spain, 27–29 August 2014; Awan, I., Younas, M., Franch, X., Quer, C., Eds.; Springer International Publishing: Cham, Switzerland, 2014; pp. 153–164.
49. D'Haro, L.F.; Kim, S.; Yeo, K.H.; Jiang, R.; Niculescu, A.I.; Banchs, R.E.; Li, H. CLARA: A Multifunctional Virtual Agent for Conference Support and Touristic Information. In *Natural Language Dialog Systems and Intelligent Assistants*; Lee, G.G., Kim, H.K., Jeong, M., Kim, J.-H., Eds.; Springer International Publishing: Cham, Switzerland, 2015; pp. 233–239. ISBN 978-3-319-19291-8.
50. de Kervenoael, R.; Hasan, R.; Schwob, A.; Goh, E. Leveraging human-robot interaction in hospitality services: Incorporating the role of perceived value, empathy, and information sharing into visitors' intentions to use social robots. *Tour. Manag.* **2020**, *78*, 104042. [CrossRef]
51. Cross, E.S.; Hortensius, R.; Wykowska, A. From social brains to social robots: Applying neurocognitive insights to human–robot interaction. *Philos. Trans. R. Soc. B Biol. Sci.* **2019**, *374*, 20180024. [CrossRef]
52. Abdelrahman, Y.; Velloso, E.; Dingler, T.; Schmidt, A.; Vetere, F. Cognitive heat: Exploring the usage of thermal imaging to unobtrusively estimate cognitive load. *Proc. ACM Interact. Mob. Wearable Ubiquitous Technol.* **2017**, *1*, 1–20. [CrossRef]
53. Salazar-López, E.; Domínguez, E.; Juárez Ramos, V.; de la Fuente, J.; Meins, A.; Iborra, O.; Gálvez, G.; Rodríguez-Artacho, M.A.; Gómez-Milán, E. The mental and subjective skin: Emotion, empathy, feelings and thermography. *Conscious. Cogn.* **2015**, *34*, 149–162. [CrossRef] [PubMed]
54. Pavlidis, I.; Tsiamyrtzis, P.; Shastri, D.; Wesley, A.; Zhou, Y.; Lindner, P.; Buddharaju, P.; Joseph, R.; Mandapati, A.; Dunkin, B. Fast by nature-how stress patterns define human experience and performance in dexterous tasks. *Sci. Rep.* **2012**, *2*, 305. [CrossRef] [PubMed]
55. Ioannou, S.; Gallese, V.; Merla, A. Thermal infrared imaging in psychophysiology: Potentialities and limits. *Psychophysiology* **2014**, *51*, 951–963. [CrossRef] [PubMed]
56. Cruz-Albarran, I.A.; Benitez-Rangel, J.P.; Osornio-Rios, R.A.; Morales-Hernandez, L.A. Human emotions detection based on a smart-thermal system of thermographic images. *Infrared Phys. Technol.* **2017**, *81*, 250–261. [CrossRef]
57. Goulart, C.; Valadão, C.; Delisle-Rodriguez, D.; Caldeira, E.; Bastos, T. Emotion analysis in children through facial emissivity of infrared thermal imaging. *PLoS ONE* **2019**, *14*. [CrossRef] [PubMed]
58. Mehrabian, A. Pleasure-arousal-dominance: A general framework for describing and measuring individual differences in temperament. *Curr. Psychol.* **1996**, *14*, 261–292. [CrossRef]
59. Posner, J.; Russell, J.A.; Peterson, B.S. The circumplex model of affect: An integrative approach to affective neuroscience, cognitive development, and psychopathology. *Dev. Psychopathol.* **2005**, *17*, 715–734. [CrossRef]
60. Zhong, K.; Qiao, T.; Zhang, L. A Study of Emotional Communication of Emoticon Based on Russell's Circumplex Model of Affect. In Proceedings of the Design, User Experience, and Usability Design Philosophy and Theory, Orlando, FL, US, 26–31 July 2019; Marcus, A., Wang, W., Eds.; Springer International Publishing: Cham, Switzerland, 2019; pp. 577–596.
61. Landowska, A. Towards new mappings between emotion representation models. *Appl. Sci.* **2018**, *8*, 274. [CrossRef]
62. Mikels, J.A.; Fredrickson, B.L.; Larkin, G.R.; Lindberg, C.M.; Maglio, S.J.; Reuter-Lorenz, P.A. Emotional category data on images from the International Affective Picture System. *Behav. Res. Methods* **2005**, *37*, 626–630. [CrossRef]

63. Kosonogov, V.; Zorzi, L.D.; Honoré, J.; Martínez-Velázquez, E.S.; Nandrino, J.-L.; Martinez-Selva, J.M.; Sequeira, H. Facial thermal variations: A new marker of emotional arousal. *PLoS ONE* **2017**, *12*, e0183592. [CrossRef]
64. Diaz-Piedra, C.; Gomez-Milan, E.; Di Stasi, L.L. Nasal skin temperature reveals changes in arousal levels due to time on task: An experimental thermal infrared imaging study. *Appl. Ergon.* **2019**, *81*, 102870. [CrossRef]
65. Bando, S.; Oiwa, K.; Nozawa, A. Evaluation of dynamics of forehead skin temperature under induced drowsiness. *IEEJ Trans. Electr. Electron. Eng.* **2017**, *12*, S104–S109. [CrossRef]
66. Liapis, A.; Katsanos, C.; Sotiropoulos, D.G.; Karousos, N.; Xenos, M. Stress in interactive applications: Analysis of the valence-arousal space based on physiological signals and self-reported data. *Multimed. Tools Appl.* **2017**, *76*, 5051–5071. [CrossRef]
67. Ioannou, S.; Ebisch, S.; Aureli, T.; Bafunno, D.; Ioannides, H.A.; Cardone, D.; Manini, B.; Romani, G.L.; Gallese, V.; Merla, A. The Autonomic Signature of Guilt in Children: A Thermal Infrared Imaging Study. *PLoS ONE* **2013**, *8*, e79440. [CrossRef]
68. Manini, B.; Cardone, D.; Ebisch, S.; Bafunno, D.; Aureli, T.; Merla, A. Mom feels what her child feels: Thermal signatures of vicarious autonomic response while watching children in a stressful situation. *Front. Hum. Neurosci.* **2013**, *7*. [CrossRef]
69. Garbey, M.; Sun, N.; Merla, A.; Pavlidis, I. Contact-Free Measurement of Cardiac Pulse Based on the Analysis of Thermal Imagery. *IEEE Trans. Biomed. Eng.* **2007**, *54*, 1418–1426. [CrossRef]
70. Lewis, G.F.; Gatto, R.G.; Porges, S.W. A novel method for extracting respiration rate and relative tidal volume from infrared thermography. *Psychophysiology* **2011**, *48*, 877–887. [CrossRef]
71. Murthy, R.; Pavlidis, I. Noncontact measurement of breathing function. *IEEE Eng. Med. Biol. Mag.* **2006**, *25*, 57–67. [CrossRef] [PubMed]
72. Pereira, C.B.; Yu, X.; Czaplik, M.; Rossaint, R.; Blazek, V.; Leonhardt, S. Remote monitoring of breathing dynamics using infrared thermography. *Biomed. Opt. Express* **2015**, *6*, 4378–4394. [CrossRef] [PubMed]
73. Fei, J.; Pavlidis, I. Thermistor at a distance: Unobtrusive measurement of breathing. *IEEE Trans. Biomed. Eng.* **2009**, *57*, 988–998. [PubMed]
74. Abbas, A.K.; Heimann, K.; Jergus, K.; Orlikowsky, T.; Leonhardt, S. Neonatal non-contact respiratory monitoring based on real-time infrared thermography. *Biomed. Eng. OnLine* **2011**, *10*, 93. [CrossRef] [PubMed]
75. Kiashari, S.E.H.; Nahvi, A.; Homayounfard, A.; Bakhoda, H. Monitoring the Variation in Driver Respiration Rate from Wakefulness to Drowsiness: A Non-Intrusive Method for Drowsiness Detection Using Thermal Imaging. *J. Sleep Sci.* **2018**, *3*, 1–9.
76. Cho, Y.; Bianchi-Berthouze, N. Physiological and Affective Computing through Thermal Imaging: A Survey. *ArXiv* **2019**, *1908*, 10307.
77. Thermography Guidelines. Standards and Protocols. Available online: http://www.iact-org.org/professionals/thermog-guidelines.html (accessed on 18 February 2020).
78. Cho, Y.; Julier, S.J.; Bianchi-Berthouze, N. Instant Stress: Detection of Perceived Mental Stress Through Smartphone Photoplethysmography and Thermal Imaging. *JMIR Ment. Health* **2019**, *6*, e10140. [CrossRef]
79. Cho, Y. Automated mental stress recognition through mobile thermal imaging. In Proceedings of the 2017 Seventh International Conference on Affective Computing and Intelligent Interaction (ACII), San Antonio, TX, USA, 23–26 October 2017; pp. 596–600.
80. Cho, Y.; Bianchi-Berthouze, N.; Julier, S.J. DeepBreath: Deep learning of breathing patterns for automatic stress recognition using low-cost thermal imaging in unconstrained settings. In Proceedings of the 2017 Seventh International Conference on Affective Computing and Intelligent Interaction (ACII), San Antonio, TX, USA, 23–26 October 2017; pp. 456–463.
81. Basu, A.; Dasgupta, A.; Thyagharajan, A.; Routray, A.; Guha, R.; Mitra, P. A Portable Personality Recognizer Based on Affective State Classification Using Spectral Fusion of Features. *IEEE Trans. Affect. Comput.* **2018**, *9*, 330–342. [CrossRef]
82. Vinciguerra, V.; Ambra, E.; Maddiona, L.; Romeo, M.; Mazzillo, M.; Rundo, F.; Fallica, G.; di Pompeo, F.; Chiarelli, A.M.; Zappasodi, F. PPG/ECG multisite combo system based on SiPM technology. In Proceedings of the Convegno Nazionale Sensori, Catania, Italy, 21–23 February 2018; Springer: Berlin/Heidelberg, Germany, 2018; pp. 353–360.

83. Chiarelli, A.M.; Bianco, F.; Perpetuini, D.; Bucciarelli, V.; Filippini, C.; Cardone, D.; Zappasodi, F.; Gallina, S.; Merla, A. Data-driven assessment of cardiovascular ageing through multisite photoplethysmography and electrocardiography. *Med. Eng. Phys.* **2019**, *73*, 39–50. [CrossRef]
84. Ruminski, J.; Kwasniewska, A. Evaluation of respiration rate using thermal imaging in mobile conditions. In *Application of Infrared to Biomedical Sciences*; Springer: Berlin/Heidelberg, Germany, 2017; pp. 311–346.
85. Cho, Y.; Julier, S.J.; Marquardt, N.; Bianchi-Berthouze, N. Robust tracking of respiratory rate in high-dynamic range scenes using mobile thermal imaging. *Biomed. Opt. Express* **2017**, *8*, 4480–4503. [CrossRef]
86. Goulart, C.; Valadão, C.; Delisle-Rodriguez, D.; Funayama, D.; Favarato, A.; Baldo, G.; Binotte, V.; Caldeira, E.; Bastos-Filho, T. Visual and Thermal Image Processing for Facial Specific Landmark Detection to Infer Emotions in a Child-Robot Interaction. *Sensors* **2019**, *19*, 2844. [CrossRef]
87. Filippini, C.; Spadolini, E.; Cardone, D.; Merla, A. Thermal Imaging Based Affective Computing for Educational Robot. In Proceedings of the Multidisciplinary Digital Publishing Institute Proceedings, Florence, Italy, 16–19 September 2019; Volume 27, p. 27.
88. Merla, A. Thermal expression of intersubjectivity offers new possibilities to human–machine and technologically mediated interactions. *Front. Psychol.* **2014**, *5*. [CrossRef]
89. Sorostinean, M.; Ferland, F.; Tapus, A. Reliable stress measurement using face temperature variation with a thermal camera in human-robot interaction. In Proceedings of the 2015 IEEE-RAS 15th International Conference on Humanoid Robots (Humanoids), Seoul, Korea, 3–5 November 2015; pp. 14–19.
90. Agrigoroaie, R.; Tapus, A. Detecting Deception in a Human-Robot Interaction Scenario. In Proceedings of the Companion of the 2017 ACM/IEEE International Conference on Human-Robot Interaction, Vienna, Austria, 6–9 March 2017; pp. 59–60.
91. Boccanfuso, L.; Wang, Q.; Leite, I.; Li, B.; Torres, C.; Chen, L.; Salomons, N.; Foster, C.; Barney, E.; Ahn, Y.A. A thermal emotion classifier for improved human-robot interaction. In Proceedings of the 2016 25th IEEE International Symposium on Robot and Human Interactive Communication (RO-MAN), New York, NY, USA, 26–31 August 2016; pp. 718–723.
92. Moliné, A.; Dominguez, E.; Salazar-López, E.; Gálvez-García, G.; Fernández-Gómez, J.; De la Fuente, J.; Iborra, O.; Tornay, F.J.; Milán, E.G. The mental nose and the Pinocchio effect: Thermography, planning, anxiety, and lies. *J. Investig. Psychol. Offender Profiling* **2018**, *15*, 234–248. [CrossRef]
93. Moliné, A.; Gálvez-García, G.; Fernández-Gómez, J.; De la Fuente, J.; Iborra, O.; Tornay, F.; Martín, J.L.M.; Puertollano, M.; Milán, E.G. The Pinocchio effect and the Cold Stress Test: Lies and thermography. *Psychophysiology* **2017**, *54*, 1621–1631. [CrossRef] [PubMed]
94. Merla, A.; Romani, G.L. Thermal Signatures of Emotional Arousal: A Functional Infrared Imaging Study. In Proceedings of the 2007 29th Annual International Conference of the IEEE Engineering in Medicine and Biology Society, Lyon, France, 22–26 August 2007; pp. 247–249.
95. Porges, S.W. The Polyvagal Theory: Phylogenetic contributions to social behavior. *Physiol. Behav.* **2003**, *79*, 503–513. [CrossRef]
96. Scassellati, B.; Brawer, J.; Tsui, K.; Nasihati Gilani, S.; Malzkuhn, M.; Manini, B.; Stone, A.; Kartheiser, G.; Merla, A.; Shapiro, A.; et al. Teaching Language to Deaf Infants with a Robot and a Virtual Human. In Proceedings of the 2018 CHI Conference on Human Factors in Computing Systems, Montreal QC, Canada, 21–26 April 2018; ACM: New York, NY, USA, 2018.
97. Petitto, L.A.; Berens, M.S.; Kovelman, I.; Dubins, M.H.; Jasinska, K.; Shalinsky, M. The "Perceptual Wedge Hypothesis" as the basis for bilingual babies' phonetic processing advantage: New insights from fNIRS brain imaging. *Brain Lang.* **2012**, *121*, 130–143. [CrossRef] [PubMed]
98. Nasihati Gilani, S.; Traum, D.; Merla, A.; Hee, E.; Walker, Z.; Manini, B.; Gallagher, G.; Petitto, L.-A. Multimodal Dialogue Management for Multiparty Interaction with Infants. In Proceedings of the 20th ACM International Conference on Multimodal Interaction, Boulder, CO, USA, 16–20 October 2018; ACM: New York, NY, USA, 2018; pp. 5–13.
99. Nasihati Gilani, S.; Traum, D.; Sortino, R.; Gallagher, G.; Aaron-lozano, K.; Padilla, C.; Shapiro, A.; Lamberton, J.; Petitto, L. Can a Virtual Human Facilitate Language Learning in a Young Baby? In Proceedings of the 18th International Conference on Autonomous Agents and MultiAgent Systems, Montreal, QC, Canada, 13–17 May 2019; International Foundation for Autonomous Agents and Multiagent Systems: Richland, SC, USA, 2019; pp. 2135–2137.

100. Nasihati Gilani, S.; Traum, D.; Sortino, R.; Gallagher, G.; Aaron-Lozano, K.; Padilla, C.; Shapiro, A.; Lamberton, J.; Petitto, L.-A. Can a Signing Virtual Human Engage a Baby's Attention? In Proceedings of the 19th ACM International Conference on Intelligent Virtual Agents, Paris, France, 2'5 July 2019; ACM: New York, NY, USA, 2019; pp. 162–169.
101. Petitto, L.-A. Hearing Babies Respond to Language's Patterning and Socially-Contingent Interactions with a Signing Avatar: Insights into Human Language Acquisition. Available online: https://www.petitto.net/published-refereed-abstract-confere (accessed on 2 March 2020).
102. Essa, I.A.; Pentland, A.P. Coding, analysis, interpretation, and recognition of facial expressions. *IEEE Trans. Pattern Anal. Mach. Intell.* **1997**, *19*, 757–763. [CrossRef]
103. Breazeal, C.; Aryananda, L. Recognition of Affective Communicative Intent in Robot-Directed Speech. *Auton. Robots* **2002**, *12*, 83–104. [CrossRef]
104. Busso, C.; Deng, Z.; Yildirim, S.; Bulut, M.; Lee, C.M.; Kazemzadeh, A.; Lee, S.; Neumann, U.; Narayanan, S. Analysis of emotion recognition using facial expressions, speech and multimodal information. In Proceedings of the 6th international conference on Multimodal interfaces, State College, PA, USA, 14–15 October 2004; Association for Computing Machinery: New York, NY, USA, 2004; pp. 205–211.
105. Lee, C.M.; Yildirim, S.; Bulut, M.; Kazemzadeh, A.; Busso, C.; Deng, Z.; Lee, S.; Narayanan, S. Emotion Recognition Based on Phoneme Classes 4. Available online: https://www.isca-speech.org/archive/interspeech_2004/i04_0889.html (accessed on 20 February 2020).
106. Tin, L.N.; Foo, S.W.; De Silva, L.C. Speech Based Emotion Classification. In Proceedings of the IEEE Region 10 International Conference on Electrical and Electronic Technology. TENCON 2001, Singapore, 19–22 August 2001; (Cat. No. 01CH37239). IEEE: Piscataway, NJ, USA; Volume 1, pp. 297–301. Available online: https://ieeexplore.ieee.org/abstract/document/949600 (accessed on 22 February 2020).
107. Stemberger, J.; Allison, R.S.; Schnell, T. Thermal imaging as a way to classify cognitive workload. In Proceedings of the 2010 Canadian Conference on Computer and Robot Vision, Ottawa, ON, Canada, 31 May–2 June 2010; pp. 231–238.
108. Or, C.K.; Duffy, V.G. Development of a facial skin temperature-based methodology for non-intrusive mental workload measurement. *Occup. Ergon.* **2007**, *7*, 83–94.
109. Lohani, M.; Payne, B.R.; Strayer, D.L. A review of psychophysiological measures to assess cognitive states in real-world driving. *Front. Hum. Neurosci.* **2019**, *13*. [CrossRef]
110. Kajiwara, S. Evaluation of driver's mental workload by facial temperature and electrodermal activity under simulated driving conditions. *Int. J. Automot. Technol.* **2014**, *15*, 65–70. [CrossRef]
111. Kiashari, S.E.H.; Nahvi, A.; Bakhoda, H.; Homayounfard, A.; Tashakori, M. Evaluation of driver drowsiness using respiration analysis by thermal imaging on a driving simulator. *Multimed. Tools Appl.* **2020**, 1–23. [CrossRef]
112. Abouelenien, M.; Burzo, M. Detecting Thermal Discomfort of Drivers Using Physiological Sensors and Thermal Imaging. *IEEE Intell. Syst.* **2019**, *34*, 3–13. [CrossRef]
113. Agrawal, K.; Subramanian, A. Enhancing Object Detection in Adverse Conditions using Thermal Imaging. *ArXiv* **2019**, *1909*, 13551.
114. Miethig, B.; Liu, A.; Habibi, S.; Mohrenschildt, M.V. Leveraging thermal imaging for autonomous driving. In Proceedings of the 2019 IEEE Transportation Electrification Conference and Expo (ITEC), Detroit, MI, USA, 19–21 June 2019; pp. 1–5.
115. Koolagudi, S.G.; Rao, K.S. Emotion recognition from speech: A review. *Int. J. Speech Technol.* **2012**, *15*, 99–117. [CrossRef]
116. Ekman, P. An argument for basic emotions. *Cogn. Emot.* **1992**, *6*, 169–200. [CrossRef]
117. Russel, J. Three dimensions of emotion. *J. Soc. Psychol.* **1980**, *9*, 1161–1178.
118. Kiebel, S.; Holmes, A.P. The general linear model. *Hum. Brain Funct.* **2003**, *2*, 725–760.
119. Chiarelli, A.M.; Romani, G.L.; Merla, A. Fast optical signals in the sensorimotor cortex: General Linear Convolution Model applied to multiple source–detector distance-based data. *NeuroImage* **2014**, *85*, 245–254. [CrossRef]
120. Suykens, J.A.; Vandewalle, J. Least squares support vector machine classifiers. *Neural Process. Lett.* **1999**, *9*, 293–300. [CrossRef]

121. LeCun, Y.; Bengio, Y.; Hinton, G. Deep learning. *Nature* **2015**, *521*, 436–444. [CrossRef]
122. Croce, P.; Zappasodi, F.; Marzetti, L.; Merla, A.; Pizzella, V.; Chiarelli, A.M. Deep Convolutional Neural Networks for feature-less automatic classification of Independent Components in multi-channel electrophysiological brain recordings. *IEEE Trans. Biomed. Eng.* **2018**, *66*, 2372–2380. [CrossRef]

applied sciences

MDPI

Article

Call Redistribution for a Call Center Based on Speech Emotion Recognition

Milana Bojanić [1,*], Vlado Delić [1] and Alexey Karpov [2]

[1] Faculty of Technical Sciences, University of Novi Sad, 21000 Novi Sad, Serbia; vlado.delic@uns.ac.rs
[2] St. Petersburg Institute for Informatics and Automation of the Russian Academy of Sciences, SPIIRAS, 14th Line 39, 199178 St. Petersburg, Russia; karpov@iias.spb.su
* Correspondence: milana.bojanic@uns.ac.rs

check for updates

Received: 26 April 2020; Accepted: 25 June 2020; Published: 6 July 2020

Featured Application: A call redistribution method for a call center based on speech emotion recognition is proposed. The research goal is efficiency improvement in emergency call centers based on automatic recognition of more urgent callers.

Abstract: Call center operators communicate with callers in different emotional states (anger, anxiety, fear, stress, joy, etc.). Sometimes a number of calls coming in a short period of time have to be answered and processed. In the moments when all call center operators are busy, the system puts that call on hold, regardless of its urgency. This research aims to improve the functionality of call centers by recognition of call urgency and redistribution of calls in a queue. It could be beneficial for call centers giving health care support for elderly people and emergency call centers. The proposed recognition of call urgency and consequent call ranking and redistribution is based on emotion recognition in speech, giving greater priority to calls featuring emotions such as fear, anger and sadness, and less priority to calls featuring neutral speech and happiness. Experimental results, obtained in a simulated call center, show a significant reduction in waiting time for calls estimated as more urgent, especially the calls featuring the emotions of fear and anger.

Keywords: emotion recognition; intelligent speech signal processing; affective computing; human computer interaction; supervised learning

1. Introduction

Spoken language processing combines knowledge from the interdisciplinary area of natural language processing, cognitive sciences, dialogue systems, and information access. Speech Emotion Recognition (SER) and text-to-speech synthesis (TTS), including voice and style conversion, as part of human–machine spoken dialogue systems correspond to certain cognitive aspects underlying the human language processing system [1]. In the last few decades, there has been growing interest in developing human–machine interfaces that are more adaptive and responsive to a user's behavior [2]. In that sense, the use of emotion in speech synthesis and recognition of emotion in speech takes an important place in attempts to improve naturalness of human–machine interaction (HMI) [3]. As to TTS, different applications such as smart environments, virtual assistants, intelligent robots, and call centers have set requirements for different speech styles identified with corresponding emotional expressions [4]. Recognition of emotions in HMI is not restricted to speech analysis only, but also image analysis (facial expression recognition, eye-tracking data) and physiological signals (pulse rate, skin conductance, facial electromyography, electroencephalography (EEG) signal) [5]. Emotion recognition in spoken dialogue systems such as call centers provides a possibility to respond to callers according to the detected emotional state or to pass control over to human operators [2,6–8].

In the SER research, two main approaches are utilized in describing the emotional space. The first approach describes the emotional space with a finite number of prototypical emotions according with categorical emotion model. The second approach uses dimensions (typically arousal and valence) to determine possible emotional states in the space defined by chosen dimensions. The latter approach corresponds to dimensional emotion models. Dimensional emotion models mostly use two or three dimensions (e.g., valence, arousal, and sometimes dominance) to describe the emotional space in which the emotional variability is to the greatest extent determined by the first two dimensions and thus used as a basis for research in the field of SER [9]. The valence dimension describes the pleasantness of emotion and ranges from positive (e.g., joy) to negative (e.g., anger). The arousal dimension indicates the level of activation during some emotional experience and it ranges from passive (e.g., sleepiness) to active (e.g., high excitement). The position of some basic, categorical emotions in the valence–arousal space is shown in Figure 1.

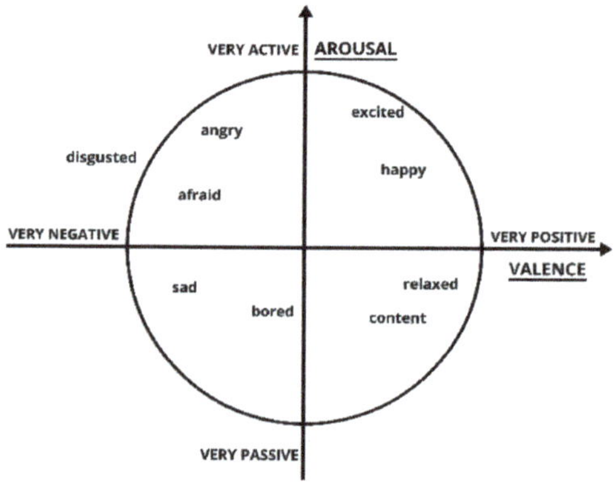

Figure 1. The circumplex model of affect in the valence–arousal space. Adapted from [10].

The dimensional models allow using emotional categories (appropriately positioned in a two-dimensional emotional space) among which it is possible to determine a distance metric [10]. Goncalves et al. utilized four dimensions (namely, valence, arousal, sense of control, and ease in achieving a goal) to describe user's emotional state while interacting with an electronic game [11]. Landowska proposed a procedure to obtain new mappings with mapping matrices for estimating the dimensions of a valence-arousal-dominance model from Ekman's six basic emotions [12]. The procedure, as well as the proposed metrics, might be used, not only in evaluation of the mappings between representation models, but also in a comparison of emotion recognition and annotation results. Emotion valence classification in self-assessed affect challenge is reported in [13]. Detection of a degree of speaker's sleepiness can help recognizing his/her emotional arousal as well [14]. Sometimes both approaches, emotion category and valence-arousal classification, are utilized for comparison, as in the INTERSPEECH Emotion Sub-Challenge on acted speech corpus [15].

In a situation when all call center operators are busy and unable to answer a new call, the system puts that call on hold regardless of its urgency. By way of illustration, if a call is the fifth call in the queue in a given moment, a caller which is terrified, angry, or upset would be left to wait for a certain period of time before his/her call is considered. This period of time is equivalent to the time in which all the preceding calls are answered. This classical approach in call centers does not take into consideration the urgency of a call and calls are processed in the order in which they are received. Petrushin utilized the emotion recognition as a part of a decision support system for prioritizing telephone voice messages in a call center and assigning a proper agent to respond the message [6]. His goal was to recognize two possible states: "agitation" which includes anger, happiness and fear, and "calm" which includes normal state and sadness. The average recognition accuracy was in the range of 73–77%.

In this research, the first presumption was that there are some calls which are more urgent and which should be processed faster. The second presumption was that the urgency of the call correlates with a caller emotional state reflected through speech. The motivation behind this research was to improve the effectiveness of call center service through giving the first level priority to the callers who are experiencing a negative valence emotional state (fear and anger), the second level priority to a sad or neutral emotional state, and the third level priority to a joyful emotional state. The proposed approach consists of recognition of caller's emotional speech and redistribution of the calls according to the proposed emotion ranking. Thus, faster processing and the decrease in waiting time for callers estimated as more urgent, is achieved.

The paper is organized as follows: Section 2 covers related works including acoustic modeling of emotional speech and the underlying emotional speech corpus, as well as methods for emotion classification. The proposed algorithm for redistribution of calls is described in detail in Section 3. The simulation and experimental results are reported and discussed in Section 4. Finally, conclusion remarks and future research directions are summarized in Section 5.

2. Materials and Methods

2.1. Emotional Speech Corpus

The GEES (Corpus of Verbal Expressions of Emotions and Attitudes—in Serbian: *Korpus Govorne Ekspresije Emocija i Stavova*) is the first corpus of acted emotional speech recorded in Serbian [16]. Six actors (3 female, 3 male) were recorded while verbally expressing semantically neutral textual material into five basic emotions: anger, joy, fear, sadness, and neutral. The underlying textual material included 32 isolated words, 30 short sentences, 30 long sentences, and one passage of 79 words. The corpus was evaluated by human listeners and reported recognition accuracy was 94.7% [16]. In this study, a part of corpus containing short and long sentences was taken into consideration because it better reflects a real conversation scenario. The isolated words and the passage were omitted from the research. Aiming to have each speaker equally represented, 58 recorded utterances from every speaker in each emotion class were used for the feature extraction. The total number of utterances used in experiments was 1740. It has been pointed out that acted emotions are more exaggerated than real ones [17] and discussed that acted emotions have limited application in real-life situations. Still, by studying the acoustic features of emotional speech on the acted emotion corpus, one can analyze acoustic variations and get insight into acoustic correlates of emotional speech. Those acoustic correlates are, to a greater extent, present in emotions occurring in real life situations or in elicited emotional speech. In that sense, the relationships between the acoustic features and the acted emotions, as well as between the acoustic features and the real life emotions, do not contradict [18]. Using acted emotions in emotional speech recognition is a way to obtain and study generic (maybe universal) expressions of emotions [19]. Additionally, our research setting was to recognize more intensive emotional states which are reflecting more urgent callers. These intensive vocal emotional expressions are more frequent in acted emotional speech corpora than in natural speech corpora.

2.2. Acoustic Modeling

The most commonly used acoustic features for SER are: prosodic features (pitch, intensity, duration), cepstral features (MFCC), spectral features (formant position and bandwidth), and occasionally voice quality features (harmonic-to-noise ratio, jitter, shimmer), in line with the studies [19–23]. The task of finding a robust feature set has led to the idea of applying statistical functionals to low-level descriptors (LLD) and resulted in very large feature vectors containing up to a few thousands of prosodic and spectral features [19]. Recently, new trends in machine learning have been directing research of automatic affect recognition towards end-to-end technique that combines deep, convolutional and recurrent neural networks trained directly on underlying raw audio signal [24,25]. A proposal of multilevel model based on a combination of LLDs and convolutional recurrent neural network model is given in [26]. Still, a lot of research in the area is based on hand-crafted features that have shown to be robust in many computational paralinguistics tasks such as emotion, autism, accent, addressee, deception, cognitive and physical load detection, and so on [20–22] (list of the INTERSPEECH Paralinguistics Challenge tasks up to 2019 is available at http://www.compare.openaudio.eu/tasks/). Schuller et al. introduced the INTERSPEECH 2013 ComParE feature set [20]. It contained 6373 features including energy, spectral, cepstral (MFCC) and voicing related LLDs (pitch, voicing probability, jitter, shimmer), as well as a few LLDs including logarithmic harmonic-to-noise ratio (HNR), spectral harmonicity and psychoacoustic spectral sharpness, etc. This set of hand-crafted acoustic features is still state-of-the-art now [27]. Another more minimalistic feature set proposed in [23] includes prosodic, excitation, vocal tract, and spectral descriptors, obtained by applying functionals to 18 LLDs that give a total of 52 utterance-level features only. Kaya et al. used the ComParE feature set for the proposed cascaded normalization. The proposed normalization approach, combining speaker level, value level and feature vector level normalization, has shown a superior performance in the task of cross-corpus acoustic emotion recognition on five corpora recorded in five languages [28]. Utterance level features, obtained through the statistical analysis of prosodic features (pitch, energy), spectral information (formants, spectrum centroid and spectrum cut-off frequency) and cepstral information (mel-frequency bands energy), are extracted to recognize seven basic emotions in Mandarin [29]. While in some studies SER relies on the prosodic and voice quality feature set only [30], and in others on cepstral features only [31], our previous study showed that a combination of both spectral and prosodic features has a higher discrimination capability for speech emotion recognition than prosodic or spectral features used separately [32]. Wagner et al. compared and discussed the advantages and usability of hand-crafted and learned representations (an end-to-end system that learns the data representation directly from the raw waveforms) [33]. Their research suggests that hand-crafted features can better generalize to unseen data and can also provide a better robustness to various acoustic conditions in comparison to purely end-to-end systems.

The proposed approach to acoustic modeling is based on the statistical analysis of the acoustic feature contours and it is performed in three steps. The openSMILE toolkit [34], used as official baseline for the series of INTERSPEECH Computational Paralinguistics challenges, is used to extract the acoustic feature set. The first step includes the extraction of short-term pitch, energy and 12 MFCC values on a frame basis. Additionally, the voicing probability and the zero crossing rate are calculated for every frame. Sequences of those short-term pitch, energy and MFCC values form feature contours. In the second step, the first derivative of the acoustic features is calculated in order to model the dynamics of speech parameters. The third step of the feature extraction process involves a statistical analysis of the feature contours. The proposed set of 12 statistical functionals has been chosen from three groups of functionals which are the most frequently used [19]:

1. The first four moments (mean, standard deviation, skewness and kurtosis);
2. Extrema and their positions (minimum, maximum, range, the relative position of minimum and the relative position of maximum);
3. Regression coefficients (the slope and the offset) and the mean squared regression error.

Finally, the extracted feature set results in 384 features for each of the processed utterances.

2.3. *Classification Methods*

A recent survey in the field of SER provided an overview of traditional classifiers and deep learning algorithms applied for SER [35]. Among traditional classifiers, they listed Hidden Markov Model (HMM), Gaussian Mixture Model (GMM), Support Vector Machines (SVM), and Artificial Neural Networks (ANN), Decision Trees (DT), k-Nearest Neighbor (kNN), k-means, and Naive Bayes Classifiers, concluding that there is no generally accepted machine learning algorithm used in this field. Recently, the focus on research changed direction towards Deep Neural Networks (DNN), with most widely used Convolutional Neural Networks (CNN), and Recurrent Neural Networks (RNN). For the purpose of speech emotion classification in this study, Linear Discriminant Classifiers (LDC) and kNN are taken into account due to their simplicity, efficiency and low computational requirements. LDCs and kNN classifiers have been used since the very first studies and turned out to be quite successful for both acted and spontaneous emotional speech [19]. Zbancioc et al. used a weighted kNN classifier for the classification task of four emotions (anger, sadness, joy and neutral) contained in the SROL emotion corpus utilized in their research [36].

In all our experiments, 10-fold partitioning of the data set was used to estimate the recognition accuracy of a particular classifier. Training and test sets included utterances from all six speakers, so these results belong to speaker-dependent experiments. Although "speaker-independent" experiments (e.g., leave-one-speaker-out) are possible on the GEES with 6 speakers (for example, the results reported by [37]), we decided to perform speaker-dependent tests in order to train classifier with more samples belonging to different speakers. In such a way, the acoustic variability present in the feature space is better modelled providing better prediction ability even when tested with an unknown speaker. Indeed, the accuracies obtained in speaker-independent cross-validation tend to be lower than the accuracies obtained in speaker-dependent cross-validation [37], but not significantly [38].

The first considered classifier is the linear Bayes classifier with the underlying assumption that classes are modeled by Gaussian densities and equal covariance matrices. Maximum likelihood estimates of Gaussian density parameters are used. As to the linear Bayes classifier, the average recognition accuracy achieved in our emotion classification experiments on the GEES corpus was 91.5% [32]. Joy was recognized with 84.2% and anger with 88.8% recognition rate. Class recognition rates for fear, neutral and sadness in the case of linear Bayes classifier were 92.5%, 97.1% and 94.8%, respectively. Table 1 shows a normalized confusion matrix for the linear Bayes classifier applied on the GEES corpus. From Table 1, it could be noted that sadness is misrecognized as a neutral state in 4% of test samples and fear is confused with neutral state in 3.2%. Neutral state has the highest recognition rate, thus its misclassification as fear and joy is about 1%. Anger and joy have lower recognition rates due to the problem of mutual misclassification, about 11% of anger test samples is recognized as joy and almost 15% of joy is misrecognized as fear.

Table 1. Normalized confusion matrix for linear Bayes classifier.

	Recognized Emotion Class (%)				
True Emotion Class	**Anger**	**Fear**	**Joy**	**Neutral**	**Sadness**
Anger	88.8	0	11.2	0	0
Fear	1.4	92.5	1.4	3.2	1.4
Joy	14.9	0.6	84.2	0.3	0
Neutral	0	1.1	1.2	97.1	0.6
Sadness	0	1.2	0	4	94.8

For the second classifier, the kNN classifier is used as a very intuitive method that classifies unlabeled examples based on their similarity to examples in the training set. It implicitly involves non-parametric density estimation, which leads to a very simple approximation of the linear Bayes classifier. Employing high dimensionality feature vectors, dimensionality reduction is sometimes applied in order to improve classification results, as in [39], where a speaker-penalty graph learning is proposed to penalize the impact of different speakers. Due to the fact that the recognition accuracy of the kNN classifier is affected by the high dimensionality of feature set, linear discriminant analysis (LDA) feature reduction has been applied on feature set [40]. In the five-class emotion classification task on the GEES corpus, the kNN classifier achieved the average recognition accuracy of 91.3% after LDA feature reduction and with k = 9. The lowest class recognition rate was obtained for joy (83.6%) and anger (86.8%). Regarding fear, neutral and sadness, higher class recognition rates were achieved—93.7%, 95.9% and 96.3%, respectively. Employing LDA, kNN achieved the average accuracy almost equal to the best result in our experiments (91.5%). In the case of linear Bayes classifier, there were no improvements after LDA feature reduction probably due to good linear separability between classes in the original feature space. Using both classification methods in our SER experiments, lower recognition results obtained for joy and anger may be explained with the observed tendency in human perception tests to misclassify anger and joy from the GEES corpus [16].

In our earlier study, the SER experiments on the same GEES corpus were performed using a multilayer perceptron (MLP) with one hidden layer [41]. The number of neurons in the input layer was equal to the number of extracted features (same feature vector as described in Section 2.2), and the number of neurons in the output layer was equal to the number of emotion classes (5). MLP was trained using standard backpropagation (BP) algorithm with varying number of neurons in the hidden layer. The highest recognition rate was achieved with 15 neurons in the hidden layer. Further increase in neurons in the hidden layer resulted in insignificant improvement of the recognition rate at the cost of increased computational complexity and thus longer processing time.

The average recognition accuracy achieved in emotion classification experiments with MLP was 90.4%. Joy was recognized with 82.5% and anger with 86.5% recognition rate. In the case of these two emotions, results of MLP underperform results of the linear Bayes classifier approximately by 2%. Sadness and neutral are emotions with the highest recognition rates of 97.7% and 93.7%, respectively. Fear is recognized with 91.7%.

It can be noted that two emotions with the lowest recognition rates, namely joy and anger, have a lower recognition accuracy compared to the experimental results with the linear Bayes classifier. This is an additional reason why we decided to use the linear Bayes classifier in the proposed system, besides it is a fast classification method.

2.4. Comparison of the SER Results with Other Studies

In general, it is a difficult task to objectively compare results of one SER research with other results reported in literature. This is due to a high diversity of research approaches to SER, regarding speech emotion corpora, the extracted feature set, classification methods and additional experimental settings (e.g., speaker-dependent or speaker-independent tests, cross-validation method applied). Regarding acted speech, two corpora have been used in plenty of research: Berlin Emotional Speech Database (Emo-DB) containing the total of 535 sentences uttered by 10 actors (5 male, 5 female) in seven emotional states, and Danish Emotional Speech Database (DES) containing the total of 419 utterances portrayed in five emotional states by 4 actors (2 male, 2 female) [28,37]. In the research by Hassan et al. the proposed 3DEC classification was tested on all three corpora (Emo-DB, DES, GEES), and the best results were achieved on the GEES corpus [37]. It can be explained by the fact that the GEES contains more samples available for training the classifier than other two corpora, and by the fact that the overall human recognition accuracy reported for GEES is 94.7%, against 86.1% for Emo-DB and 67.3% for DES. The overall human accuracy reflects the distinction degree of the acoustic representation of basic emotions in a corpus, which is very high for the GEES corpus.

In our earlier study [42], the comparison of basic emotion classification in valence-arousal space was made on the Emo-DB and the GEES corpora. The mapping of basic emotions into three classes along the valence axis (positive, neutral, and negative), and three classes along the arousal axis (high, neutral, and low) was performed. The recognition results along the arousal axis were above 90% for both corpora. The average recognition results along the valence axis were 83.2% for the GEES and 76.9% for Emo-DB. It is in line with the findings showing that arousal discrimination tasks, based on acoustic features, achieve higher recognition rates than valence discrimination tasks [28].

We consider the GEES with 1740 utterances portrayed in 5 emotional states by 6 actors as a suitable and adequate basis for SER research. Also, taking into account that Serbian is still an under-resourced language, there are far less available emotional speech data and the corresponding research for Serbian (GEES is the only one emotional speech corpus accessible for research purposes) even in comparison with other Slavic languages like Russian [43], Czech [44], etc.

A comparative analysis of our results and the results of some other SER studies conducted on the GEES corpus was performed. Due to the fact that this is a rather small corpus in Serbian, it was not a subject of much research. Two SER researches on the GEES corpus were found to be compared with our results.

The first study for comparison is by Hasan et al. [37], who proposed a hierarchical classification technique using SVM for binary emotion classification on every level. As to feature extraction they decided to apply a "brute force" approach and 6552 acoustic features for each utterance. The extracted feature vector included 56 low-level descriptors (among which is pitch, energy, spectral energy, MFCCs) and 39 statistical functionals applied to these LLDs and their first and second derivatives. In the experiments three acted databases were used: the Danish Emotional speech (DES), the Berlin database (Emo-DB) and the Serbian GEES database, and one spontaneous database (Aibo corpus). The proposed hierarchical classification, called 3DEC, is based on input data in such a way that input data and their confusion plots determine the hierarchy of the proposed classification scheme. They used both speaker-dependent and speaker-independent approaches for SVM-based model training and testing. We present only results of speaker-dependent tests as to be able to make a comparison with our results. For the speaker-dependent test, 10-fold cross-validation for the whole corpus is applied, as in our case.

The reported [37] average recognition accuracy on the GEES corpus is 94.1%. It is achieved with the proposed 3DEC combination of SVM classifiers in the speaker-dependent test. Recognition accuracy in their research is obtained as an unweighted average accuracy (UA), i.e., accuracy per class is averaged by the total number of classes. It should be noted that in the case of the GEES corpus, UA accuracy is identical to weighted average accuracy (WA) due to equally balanced emotion classes. Comparing our result with the result of Hasan et al. [37], it can be seen that our average recognition accuracy is lower by 3%. It should be noted that our result is obtained with a significantly smaller feature vector (384 features against 6552 features in [37]). Additionally, the classification methods applied are different. In our experiments, the linear Bayes classifier is used as a simple and fast method for training and test stages, and their proposed 3DEC combination of SVMs requires training of four SVMs. We consider that our proposed SER achieves a slightly lower result compared to the best recognition accuracy reported for the GEES (94.1% in [37]), but having significantly smaller feature vectors and computationally less demanding classification method.

One more study on the GEES corpus, by Shaukat et al. [45], applied the multistage (hierarchical) emotion categorization with SVM. In their research, the extracted utterance-level vectors of 318 features, among which are pitch, energy, MFCC, formants and their statistical functionals (e.g., mean, variance, maximum, minimum, etc.). In the experiment on the GEES corpus, they reported the average emotion recognition rate of 90.63%.

Comparing our result with the result of Shaukat et al. [45], it can be seen that our average recognition accuracy is higher by 1%. They applied a hierarchical classification techniques with 4 SVMs, thus training of all 4 SVMs is necessary. It should be noted that feature vector set used in [45] is smaller, but an important difference is that their experiments were performed on individual speaker sub-corpora and overall recognition accuracy was calculated as an average value of recognition accuracies obtained for each individual speaker. Our recognition accuracy is evaluated after 10-fold cross-validation on the whole corpus, like in the study of Hasan et al. [37], which we consider as a more objective measure of recognition performance.

3. Algorithm for Call Redistribution Based on Speech Emotion Recognition

As mentioned earlier, in this research, one presumption was that the urgency of the call correlates with the caller's emotional state reflected through speech. Our focus was on emergency call centers and health care centers for elderly people. Aiming to recognize more urgent callers among them, we have proposed the ranking of five basic emotions.

So, basic emotions with negative valence (fear, anger and sadness) reflect unpleasantness of the speaker and our presumption was that those speakers have a health, or any other, more urgent problem. On the other hand, there are positive valence emotions (e.g., joy) and neutral valence (neutral state) that are supposed to reflect less urgent speaker's state and those calls could be processed later.

The proposed ranking of five basic emotions is:

1. Fear (f)
2. Anger (a)
3. Sadness (s)
4. Neutral (n)
5. Joy (j).

In the proposed ranking, fear is put first because it is an emotion that people experience when facing a serious problem (serious injuries, heart attack, accidents, etc.). In the research conducted on the CEMO corpus containing dialogues recorded in a real-world medical call center, it was pointed out that patients had often expressed stress, pain, fear of being sick or even real panic [8]. Fear is the most common emotion in the CEMO corpus, with different levels of intensity and many variations [7]. Anger is the second negative and high arousal emotion, expressed in various stressful and disturbing situations. Sadness is in third place. It is an emotion with negative valence which is typical for elderly and lonely people. Holmen et al. reported that experiencing loneliness had a negative influence on the state of mood, so loneliness and sad mood prevailed especially among elderly subjects with cognitive difficulties [46]. Joy is in last place because it is considered to reflect full satisfaction and good mood, which are not indicators of urgent states.

The research setting is explained using an example of five calls received at the same moment—while all operators are busy. For each call, the initial part of the caller's speech is recorded. This recording is further processed and the feature vector x^i is extracted. The feature vector is forwarded to a classifier which gives one of the five emotion labels (anger, joy, fear, sadness, and neutral) to input speech. Finally, after SER, those five calls are redistributed according to the recognized emotions and the proposed emotion ranking. The proposed framework of call processing is shown in Figure 2. For example, in the scenario shown in Figure 2, the original call order was neutral, joyful, sad, afraid, and angry; after SER and proposed call redistribution, the system will firstly process the call featuring fear, then the call featuring anger, afterwards a sad caller, then neutral, and at last the call featuring joy.

The proposed algorithm, whose block diagram is shown in Figure 3, has the following steps:

1. When a call is received while all operators are busy, the system asks for the reason of the call and records the caller's speech for about 5–8 s. This recording contains about 1–2 sentences, depending on the dialogue strategy, which will be processed quickly by SER while the call is put on hold. For each recording, the feature vector x consisting of 384 features is extracted.
2. The extracted feature vector is input to our trained SER classifier. The classifier outputs one of the five emotion labels (fear, anger, sadness, neutral and joy) to the input speech. Keywords recognized by automatic speech recognition (ASR) can also be used for sentiment analysis, but it depends on both language and type of the call center.
3. If there are several calls on hold at the same time, they are redistributed based on the associated emotion label. Redistribution is done according to the introduced priority vector p:

$$p = [p_1 = f,\ p_2 = a,\ p_3 = s,\ p_4 = n,\ p_5 = j]^T, \tag{1}$$

where f represents fear, i.e., it denotes the speaker recognized as being in a state of fear, a denotes the speaker recognized as angry, s marks the speaker recognized as sad, n refers to neutral, and j to a joyful state of the speaker. The introduced priority vector, i.e., emotion ranking, represented in Equation (1), is proposed considering application in emergency call centers and health care centers for elderly people. It should be noted that the proposed algorithm is not restricted to the aforementioned priority vector only. Regarding a specific domain of application, a new emotion ranking can be adopted.

4. Calls are processed in the new order which is obtained after their emotion labeling based on SER (and ASR) and redistribution according to the proposed emotion ranking, i.e., the priority vector.

Firstly, all callers that are recognized as afraid are processed, after them angry callers and so on. In the end, the callers recognized as joyful are processed. The final goal of the redistribution is reduction in waiting time for the callers recognized as the priority. Let us denote the waiting time t_{1i} of a caller i without SER and call redistribution, where $i = 1, \ldots, C$ and C is the number of calls received at the same moment. Then, t_{2i} denotes the waiting time of a caller i after SER and call redistribution (after application of the proposed algorithm). The objective function is:

$$\max \sum_{i=1}^{C} t_{1i} - t_{2i}, \tag{2}$$

according to the priority vector p. The objective function is formulated as to maximize waiting time reduction for the callers recognized as the priority regarding the priority vector p. So, the goal of call redistribution is to maximize waiting time reduction for the caller i, if the caller i is set as priority regarding the vector p. In our experiments this is the case for the caller recognized as being afraid—fear is in first place in the priority vector p. Afterwards, the objective function maximizes the waiting time reduction for the caller recognized as being angry, since anger is in second position in the priority vector p.

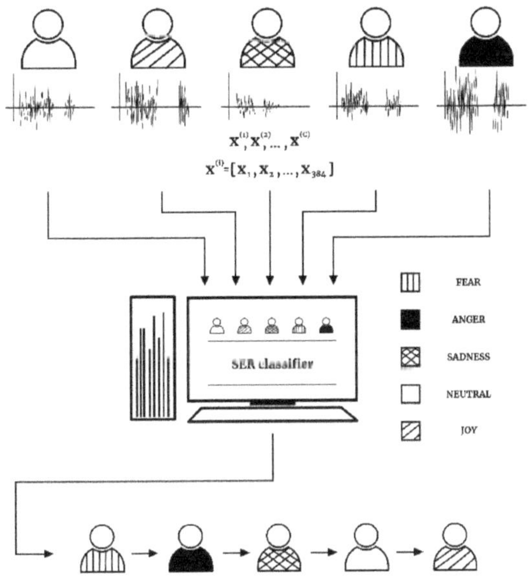

Figure 2. Proposed call redistribution based on SER.

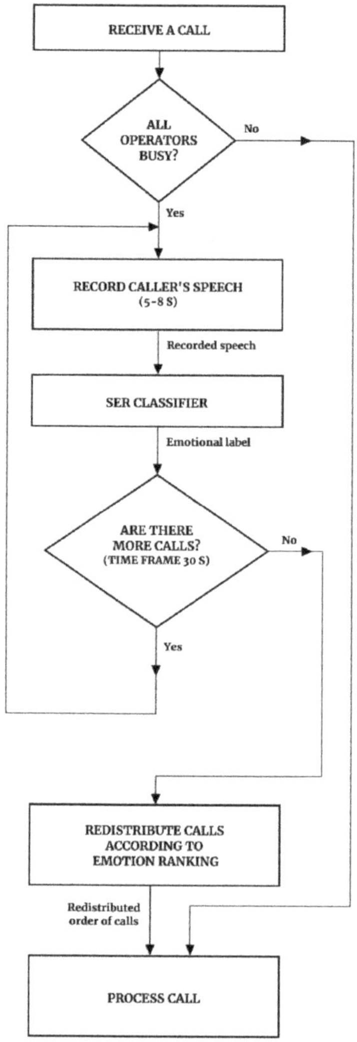

Figure 3. Block diagram of the proposed algorithm.

The call processing and the proposed algorithm are intended to be a part of a client-server application based upon computer-telephone integration (CTI). The main part of the application is running on the server side located on a remote computer. A client side is located at a call center. When a call is received, if there is at least one free operator, the call is answered immediately. In the case all operators are busy at that moment, the client side initiates a connection with the server which is waiting for clients. After the connection is established, a new session is started and the client sends a recorded speech sample of the call. On the server side, the feature vector is extracted for a received speech record and it is forwarded to SER module which classifies it into one of the predefined emotion categories. If a new call is received at a call center within a period of 30 s, the client sends the recorded speech sample of the second call and steps 1 and 2 of the algorithm are performed. The established session lasts as long as there are new calls within the time frame of 30 s, which is chosen as the overlapping time between two consecutive calls. When there are no more calls within the specified time period, all calls processed during the current session of client-server communication are redistributed according to the proposed emotion ranking. The call redistribution is intended to be applied on a finite number of calls received in a short period of time while all the operators are busy. In the experiments, the situations of three, five and seven simultaneously received calls were considered. Let us denote them as group of calls. For example, when a call center simultaneously receives seven calls, those seven calls will be redistributed according to the SER system output and processed in a new order. While operators are answering those seven calls, if there is a new incoming call, it will be put in a new group of calls to be redistributed. The proposed emotion ranking can be specified after the connection establishment, so that the server adapts the system response to the specific type of a call center (a client). At the end of the session, the server sends the client the list of redistributed calls which are then processed according to the redistributed order.

4. Simulation and Experimental Results

The research was designed as a set of experiments in a simulated call center receiving a different number of calls simultaneously, i.e., during a short period of time when all operators are busy. The experiments focused on: (i) the redistribution of calls based on emotion label assigned after speech emotion recognition task, and (ii) the evaluation of time period in which the call was put on hold without and after speech emotion recognition was applied for call redistribution. During experiments, the number of simultaneously received calls varied from 3 to 7. In all experiments, prosodic and spectral feature set was used and the linear Bayes classifier and kNN were considered as classification techniques, as described in Section 2.

An average waiting time, without and with the redistribution, for each emotional state is evaluated as an average value of waiting time obtained for 50 experimental iterations in the simulated call center with one ideal active human-operator (a human factor is not considered). This procedure is repeated for each experimental setting (3, 5, and 7 simultaneously received calls). Waiting time reduction estimate is made under assumptions about underlying distribution of emotions in input calls and distribution of call duration. We assumed that all the emotions had a uniform distribution as well as that call durations were uniformly distributed across the chosen range (from 30 s to 3 min 50 s). The specified range was chosen with the assumption that it is wide enough to take into consideration the duration of shorter, medium, and longer phone calls as well. Thus, the evaluated waiting time after call redistribution may be shorter for every caller proportionally to the number of active operators in the call center.

A pseudo-random number generator is used for the generation of emotion labels (random choice of emotion for input call) as well as the generation of input call duration. The order of the calls in queue (the order of the call arrival) has, as in simulation as in real-world call center, the biggest influence on the estimated waiting time which a caller could spend in callers' queue. In our simulations, the order of the call arrival featuring specific emotion is also unknown and thus determined by generated pseudo-random number. Thus, regarding every iteration in simulation, the random number of occurrences of each emotion class with the associated random call duration, and finally random order of calls (emotions) in callers' queue, jointly influence the variations of estimated average waiting time, before and after call redistribution. Additionally, the recognition rate of some emotional state has an influence on the average waiting time after call redistribution.

Simulation of call redistribution in a call center is explained on an experimental example for three simultaneously received calls. Each call is represented by one utterance in the GEES corpus. Firstly, the vector of randomized emotion labels for three input calls was generated. According to the input emotion label vector, three utterances belonging to chosen emotion classes were randomly (regarding a speaker) selected from the corpus and provided as an input to SER. As an initial part of the simulation, duration of a call, generated as a random value between 30 s and 3 min 50 s, was appended to each of these utterances. Knowing the initial order of the simulated calls (determined with input emotion label vector), the initial waiting time in the caller's queue is calculated for each caller as a sum of call duration for all preceding callers in the queue. Every caller is represented with input utterance determined with input emotion label. Thus, initial waiting time for every emotion class is evaluated. Secondly, based on the classifier output each input utterance gets one of the five emotion labels, thus output emotion label vector is obtained. Given the output emotion label, calls are redistributed according to the priority vector. New waiting time is calculated for each caller based on the new position in redistributed caller's queue. Accordingly, new waiting time for every emotion class is evaluated.

Table 2 shows the average waiting time which a caller will spend if his/her call is among three calls received at the same moment while all operators are busy, before and after application of SER and call redistribution. It can be observed that there is a significant waiting time reduction for callers recognized as being in a state of fear: initially, they were waiting for about 2 min 40 s, and after SER and the proposed call redistribution they had to wait only 8 s. In the case of an angry caller, there is also a noticeable waiting time reduction: the initial waiting time was 2 min 19 s and after redistribution only about 1 min. In the case of a sad caller, there is little time saving expressed in few seconds: the initial waiting time was 2 min and after redistribution reduced to 1 min 45 s. Regarding neutral and joyful emotional states of the caller, there is an increase in waiting time after SER and call redistribution: about 1 min increased waiting time for a neutral caller and about 2 min for a joyful caller. This increase was expected as the redistribution always places callers with these emotions at the end of callers' queue.

Table 2. Average waiting time when 3 calls are received simultaneously while all operators are busy.

Emotion	without the Proposed Algorithm [min]:[s]	after Application of SER and Call Redistribution [min]:[s]
fear	2:43	0:08
anger	2:19	1:01
sadness	2:00	1:45
neutral	2:09	3:07
joy	1:56	4:07

The average waiting time which a caller will spend if his/her call is among five calls received in a short period of time while all operators are busy, before and after the application of the proposed algorithm, is shown in Table 3. In the case of fear as the first in emotion ranking, there is the biggest and significant decrease in waiting time: from 4 min 17 s to 25 s after SER and redistribution. There is also a significant decrease in waiting time for angry callers: from 4 min 36 s to 1 min 57 s.

Table 3. Average waiting time when 5 calls are received simultaneously while all operators are busy.

Emotion	without the Proposed Algorithm [min]:[s]	after Application of SER and Call Redistribution [min]:[s]
fear	4:17	0:25
anger	4:36	1:57
sadness	4:49	3:56
neutral	4:07	5:13
joy	3:24	7:34

Unlike the experiment with three calls at the same time, in the experiment with five calls, callers recognized as being sad have achieved nearly 1 min shorter waiting time after SER and redistribution. In the case of a neutral state, the waiting time is increased for about 1 min. For callers recognized as being joyful, the increase is larger and amounts to about 4 min.

Table 4 shows the average waiting time which a caller will spend if his/her call is among 7 calls received simultaneously, i.e., in a short period of time while all operators are busy. As in two previous experimental settings, three emotions ranked as the priority one (fear, anger and sadness) have a significant decrease in waiting time. Calls featuring fear have the biggest waiting time reduction: it amounts to about 5 min 40 s. Calls featuring anger have achieved 2 min 20 s reduction in waiting time. In the case of a sad caller, the achieved decrease in waiting time is about 1 min. It can be observed that neutral and joyful callers have an increase in waiting time: 2 min 37 s and 5 min 30 s, respectively.

Table 4. Average waiting time when 7 calls are received simultaneously while all operators are busy.

Emotion	without the Proposed Algorithm [min]:[s]	after Application of SER and Call Redistribution [min]:[s]
fear	6:39	0:54
anger	6:33	4:11
sadness	7:16	6:24
neutral	6:12	8:49
joy	6:10	11:40

The comparative results of average waiting time in all three experimental settings (3, 5, and 7 simultaneously received calls) regarding the callers in all five emotional states, are shown in Figure 4. As the callers in the state of fear have the highest priority, their average waiting time is significantly reduced in all experimental settings, even up to twenty times in the case of three simultaneously received calls, ten times in the case of five simultaneously received calls, and six times reduced in the case of seven calls. Angry callers are given the second priority in redistribution, so in all experiments the decrease in their average waiting time is achieved. In the case of three and five simultaneously received calls, the waiting time after redistribution is reduced to less than half of the waiting time before redistribution. In the case of seven simultaneously received calls, the waiting time is reduced by one third of the initial waiting time.

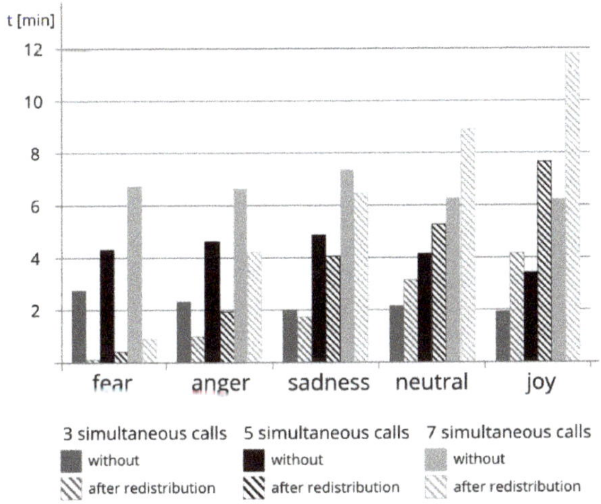

Figure 4. Average waiting time for five emotional states, without and after application of the proposed call redistribution.

From experimental results shown in Figure 4, it can be noticed that sad callers will have a moderate decrease in waiting time after the proposed call redistribution. The absolute value of waiting time reduction is the biggest in the case of seven simultaneously received calls, but the relative value of reduction is the biggest in the case of five calls and it amounts to about 18% of the initial waiting time.

As can be observed from Figure 4, callers in a neutral state have increased waiting time after call redistribution, about 1 min increase in the case of three and five simultaneously received calls, and about 2 min increase in the case of seven simultaneously received calls. Joy is marked as the emotion with the lowest priority, which is why callers featuring joy are put at the end of the caller's queue. It causes a significant increase in waiting time for the caller in a state of joy, about twice as longer waiting time after the proposed call redistribution in all experimental settings.

Table 5 shows the waiting time reduction for five emotional states after SER and the proposed call redistribution is applied, in all experimental settings (with 3, 5 and 7 simultaneously received calls) in a simulated call center. Time reduction is calculated as difference between the average waiting time without the call redistribution and the average waiting time after application of SER and the proposed call redistribution:

$$\Delta t_e = \bar{t}_{1e} - \bar{t}_{2e}, \tag{3}$$

where $\bar{t}_{1e}$ denotes the average waiting time for a caller in the emotional state e without SER and call redistribution, e denotes one of the five emotional states (fear, anger, sadness, neutral and joy), and $\bar{t}_{2e}$ denotes the average waiting time for a caller in the emotional state e after application of SER and call redistribution.

Table 5. Waiting time reduction after the proposed call redistribution is applied. Time is expressed in [min]:[s].

Emotion	3 Calls Simultaneously	5 Calls Simultaneously	7 Calls Simultaneously
fear	2:35	3:52	5:45
anger	1:18	2:39	2:22
sadness	0:15	0:53	0:52
neutral	−0:58	−1:06	−2:37
joy	−2:11	−4:10	−5:30

The positive values of waiting time reduction in Table 5 indicate the real reduction in waiting time after call redistribution, which is the case of the callers recognized as being in a state of fear, anger or sadness. Negative values of waiting time reduction indicate that waiting time after call redistribution is actually increased, which is the case of the callers recognized as being in a neutral or joyful state. From the results presented in Table 5, it can be observed that as the number of simultaneously received calls grows, the calls featuring three recognized emotions considered as indicators of more urgent caller's state, namely fear, anger and sadness, show the tendency to have a decreased waiting time after the proposed call redistribution. On the other hand, the calls featuring recognized neutral speech and joy show tendency of increased waiting time as the number of simultaneously received calls grows, but it is considered justified as long as more urgent calls are processed instead of less urgent one.

To examine the results in the case of larger number of iterations, the simulations were performed using 200, 500, and 1000 iterations in all three experimental settings (3, 5, and 7 simultaneously received calls). For each experimental setting, obtained results are presented in Tables 6–8, respectively. Regarding initial average waiting time, even with 1000 iterations there are differences in evaluated initial average waiting time across five emotional states due to combination of random order of emotions in callers' queue and random duration of each call in the queue. Similar to the experiments with 50 iterations, after application of SER and call redistribution, calls featuring fear and anger have achieved significant reduction in waiting time. Unlike the simulation with 50 iterations, calls featuring sadness achieved in some cases slight increase and in some cases slight decrease in waiting time after call redistribution. This can be explained with the fact that neutral callers are put in the middle of callers' priority, so it was expected that their waiting time after increased number of iterations is evaluated as slightly changed initial average value. As can be observed from Tables 6–8, callers recognized as being in neutral and joyful states will have increased waiting time, similar to the results obtained in the simulation with 50 iterations.

Experimental results show the decrease in waiting time of the prioritized emotions. Indeed, there is a minor probability of misrecognizing anger as joy (because both are characterized by a high arousal, but opposite valence poles), and placing that caller at the end of the callers' queue, but possible negative effect depends on the position of such a call in original queue and emotional states of other callers in it. Overall experimental results show an essential decrease in waiting time of the prioritized emotions with negative valence.

In real-world emergency call centers, it is unlikely to expect all emotions equally distributed, as it was case in our simulation experiments. It is more likely to receive more calls featuring fear and less calls featuring joy, as it is reported for the CEMO corpus recorded in a real-world medical call center [7].

Although the results of the proposed SER might be to a certain extent lower in real-world emergency call center, we consider, based on high recognition accuracy for fear, sadness, and neutral that the proposed approach to SER and call redistribution based on it would improve effectiveness of such call center service.

Table 6. Average waiting time when 3 calls are received simultaneously while all operators are busy.

	without the Proposed Algorithm [min]:[s]			**after Application of SER and Call Redistribution [min]:[s]**		
Emotion\iterations	**200**	**500**	**1000**	**200**	**500**	**1000**
fear	2:03	2:05	2:07	0:13	0:15	0:13
anger	2:10	2:00	2:13	1:12	1:21	1:06
sadness	2:08	2:02	2:17	1:49	2:14	2:13
neutral	2:14	2:07	2:11	3:14	3:07	3:09
joy	2:29	2:19	2:07	4:09	3:59	4:07

Table 7. Average waiting time when 5 calls are received simultaneously while all operators are busy.

	without the Proposed Algorithm [min]:[s]			**after Application of SER and Call Redistribution [min]:[s]**		
Emotion\iterations	**200**	**500**	**1000**	**200**	**500**	**1000**
fear	4:21	4:33	4:25	0:23	0:27	0:30
anger	4:27	4:03	4:04	2:14	2:24	2:30
sadness	4:14	4:16	4:19	4:24	4:21	4:22
neutral	4:28	4:21	4:30	6:16	6:21	6:17
joy	4:26	4:17	4:11	8:18	7:56	8:07

Table 8. Average waiting time when 7 calls are received simultaneously while all operators are busy.

	without the Proposed Algorithm [min]:[s]			**after Application of SER and Call Redistribution [min]:[s]**		
Emotion\iterations	**200**	**500**	**1000**	**200**	**500**	**1000**
fear	6:24	6:30	6:32	1:00	0:54	0:48
anger	6:18	6:35	6:22	3:30	3:40	3:42
sadness	6:41	6:09	6:30	6:31	6:32	6:26
neutral	6:12	6:34	6:29	9:07	9:19	9:22
joy	6:28	6:31	6:14	12:04	12:05	12:08

5. Conclusions

The presented research has addressed the problem occurring in emergency call centers when there are several incoming calls in a short period of time while all operators are busy. The proposed solution takes into account a caller's emotional state, by recognizing emotion in speech and giving priority to the caller with negative valence emotion (fear, anger and sadness). The research aims to improve efficiency of emergency call centers based on recognition of more urgent callers. Utilizing the proposed emotion ranking and call redistribution, there is a significant reduction in waiting time for the callers recognized as being in the state of fear. A noticeable waiting time reduction is also achieved in the case of callers recognized to be angry, and a slight reduction in the case of callers recognized to be sad. On the other hand, the algorithm puts neutral and joyful callers at the end of the call queue, so those callers will have an increased waiting time. This is the price to be paid, and it has been considered that less urgent callers are more capable of bearing a longer waiting time.

Additionally, the waiting time for the most urgent calls can be shortened by giving the signal to operators who process lower priority calls that there is an emergency call on hold. Depending on the dialogue strategy in a call center, the current call will be ended faster or put on hold, so that an emergency call would be received immediately.

Although there are evident differences between the emotional speech corpus recorded in a real call center and the acted emotional speech corpus recorded under controlled conditions, the experimental results in the simulated call center give a promising sign that the proposed approach to SER and call redistribution based on it would improve effectiveness of a real call center service. The proposed algorithm is a basis for detecting critical users in the specific type of call centers considered in the research.

Other SER techniques can be used instead of the proposed one, with similar results related to the improvement of a call center effectiveness. The proposed SER based on hand-crafted features (like at the OpenSMILE toolkit) could be faster and more robust in real conditions than any DNN or end-to-end based SER system, particularly in the case of a rather small GEES corpus, i.e., the only one available in Serbian that was suitable for the presented research. Due to the lack of available data, any DNN- or end-to-end-based SER system for Serbian could not be trained well, and there is a high risk of model over-fitting. In the only emotional speech corpus for under-resourced Serbian (GEES), there are just 1800 utterances, which is definitely not enough for state-of-the-art NN-based approaches.

Further research should consider "in the wild" recordings from real-world call centers (emergency call centers or health care centers for elderly people), so that the proposed approach could be tested on realistic data and its efficiency verified. Further research may also be directed toward combining paralinguistic and linguistic information. Recordings of the initial part of a call (1–2 sentences with duration of 5–8 s) in human–machine dialogue can be used as input not only into SER, but also into ASR. After ASR, recognized keywords can be used as an additional indicator of certain emotional states and thus priorities. It could increase reliability of the emotion estimation and utility of the proposed algorithm, even in the case of a lower arousal, i.e., more passive levels of emotion activation. Of course, a possible fusion of SER and ASR depends on the dialogue strategy, and the language and vocabulary expected in particular human–machine interactions.

Author Contributions: Conceptualization, M.B. and V.D.; methodology, M.B., V.D. and A.K.; formal analysis, M.B.; investigation, M.B.; writing—original draft preparation, M.B.; writing—review and editing, V.D. and A.K.; visualization, M.B.; supervision, V.D. and A.K. All authors have read and agreed to the published version of the manuscript.

Funding: This work has resulted from cooperation between researchers from two institutions at the project HARMONIC (ERA.Net RUS Plus, 2017-2021) related in part to human–machine interaction, as well as supported by the Russian Science Foundation project #18-11-00145 (Section 2.2).

Conflicts of Interest: The authors declare no conflict of interest.

References

1. Delić, V.; Perić, Z.; Sečujski, M.; Jakovljević, N.; Nikolić, J.; Mišković, D.; Simić, N.; Suzić, S.; Delić, T. Speech technology progress based on new machine learning paradigm. *Comput. Intel. Neurosc.* **2019**, *2019*, 4368036:1–4368036:19. [CrossRef]
2. Lee, C.M.; Narayanan, S.S. Toward detecting emotions in spoken dialogs. *IEEE Trans. Speech Audio Process.* **2005**, *13*, 293–303. [CrossRef]
3. Ten Bosch, L. Emotions, speech and ASR framework. *Speech Commun.* **2003**, *40*, 213–225. [CrossRef]
4. Suzić, S.; Delić, T.; Pekar, D.; Delić, V.; Sečujski, M. Style transplantation in neural network-based speech synthesis. *Acta Polytech. Hung.* **2019**, *16*, 171–189. [CrossRef]
5. Wrobel, M. Applicability of Emotion Recognition and Induction Methods to Study the Behavior of Programmers. *Appl. Sci.* **2018**, *8*, 323. [CrossRef]
6. Petrushin, V. Emotion in speech: Recognition and application to call centers. In Proceedings of the Conference on Artificial Neural Networks in Engineering (ANNIE), St. Louis, MO, USA, 7–10 November 1999; pp. 7–10.

7. Vidrascu, L.; Devillers, L. Five emotion classes detection in real-world call center data: The use of various types of paralinguistic features. In Proceedings of the International Workshop on Paralinguistic Speech–between Models and Data (PARALING'07), Saarbrucken, Germany, 3 August 2007; DFKI: Saarbrucken, Germany, 2007; pp. 11–16.
8. Devillers, L.; Vaudable, C.; Chastagnol, C. Real-life emotion-related states detection in call centers: A cross-corpora study. In Proceedings of the INTERSPEECH 2010, Makuhari, Chiba, Japan, 26–30 September 2010; pp. 2350–2353.
9. Nicolaou, M.A.; Gunes, H.; Pantic, M. Continuous prediction of spontaneous affect from multiple cues and modalities in valence-arousal space. *IEEE Trans. Affect. Comput.* **2011**, *2*, 92–105. [CrossRef]
10. Russell, J. A circumplex model of affect. *J. Pers. Soc. Psychol.* **1980**, *39*, 1161–1178. [CrossRef]
11. Gonçalves, V.P.; Costa, E.P.; Valejo, A.; Filho, G.; Johnson, T.M.; Pessin, G.; Ueyama, J. Enhancing intelligence in multimodal emotion assessments. *Appl. Intell.* **2017**, *46*, 470–486. [CrossRef]
12. Landowska, A. Towards New Mappings between Emotion Representation Models. *Appl. Sci.* **2018**, *8*, 274. [CrossRef]
13. Montacié, C.; Caraty, M. Vocalic, Lexical and Prosodic Cues for the INTERSPEECH 2018 Self-Assessed Affect Challenge. In Proceedings of the INTERSPEECH 2018, Hyderabad, India, 2–6 September 2018; pp. 541–545. [CrossRef]
14. Gosztolya, G. Using Fisher Vector and Bag-of-Audio-Words Representations to Identify Styrian Dialects, Sleepiness, Baby & Orca Sounds. In Proceedings of the INTERSPEECH 2019, Graz, Austria, 15–19 September 2019; pp. 2413–2417. [CrossRef]
15. Gosztolya, G.; Busa-Fekete, R.; Toth, L. Detecting Autism, Emotions and Social Signals Using AdaBoost. In Proceedings of the INTERSPEECH 2013, Lyon, France, 25–29 August 2013; pp. 220–224.
16. Jovičić, S.T.; Kašić, Z.; Djordjević, M.; Rajković, M. Serbian emotional speech database: Design, processing and evaluation. In Proceedings of the 9th International Conference Speech and Computer—SPECOM'2004, St. Petersburg, Russia, 20–22 September 2004; pp. 77–81.
17. Williams, C.; Stevens, K. Emotions and speech: Some acoustical correlates. *J. Acoust. Soc. Am.* **1972**, *52*, 1238–1250. [CrossRef] [PubMed]
18. Ayadi, M.E.; Kamel, M.S.; Karray, F. Survey on speech emotion recognition: Features, classification schemes and databases. *Pattern Recognit.* **2011**, *44*, 572–587. [CrossRef]
19. Schüller, B.; Batliner, A.; Steidl, S.; Seppi, D. Recognising realistic emotions and affect in speech: State of the art and lessons learnt from the first challenge. *Speech Commun.* **2011**, *53*, 1062–1087. [CrossRef]
20. Schuller, B.; Steidl, S.; Batliner, A.; Vinciarelli, A.; Scherer, K.; Ringeval, F.; Chetouani, M.; Weninger, F.; Eyben, F.; Marchi, E.; et al. The INTERSPEECH 2013 Computational Paralinguistics Challenge: Social Signals, Conflict, Emotion, Autism. In Proceedings of the INTERSPEECH 2013, Lyon, France, 25–29 August 2013; pp. 148–152.
21. Schuller, B.; Steidl, S.; Batliner, A.; Epps, J.; Eyben, F.; Ringeval, F.; Marchi, E.; Zhang, Y. The INTERSPEECH 2014 Computational Paralinguistics Challenge: Cognitive & Physical Load. In Proceedings of the INTERSPEECH 2014, Singapore, 14–18 September 2014; pp. 427–431.
22. Schuller, B.; Steidl, S.; Batliner, A.; Bergelson, E.; Krajewski, J.; Janott, C.; Amatuni, A.; Casillas, M.; Seidl, A.; Soderstrom, M.; et al. The INTERSPEECH 2017 Computational Paralinguistics Challenge: Addressee, Cold & Snoring. In Proceedings of the INTERSPEECH 2017, Stockholm, Sweden, 20–24 August 2017; pp. 3442–3446. [CrossRef]
23. Eyben, F.; Scherer, K.R.; Schüller, B.W.; Sundberg, J.; Andre, E.; Busso, C.; Devillers, L.Y.; Epps, J.; Laukka, P.; Narayanan, S.S.; et al. The Geneva Minimalistic Acoustic Parameter Set (GeMAPS) for voice research and affective computing. *IEEE Trans. Affect. Comput* **2016**, *7*, 190–202. [CrossRef]
24. Trigeorgis, G.; Ringeval, F.; Brueckner, R.; Marchi, E.; Nicolaou, M.A.; Schuller, B.; Zafeiriou, S. Adieu features? End-to-end speech emotion recognition using a deep convolutional recurrent network. In Proceedings of the 41st IEEE International Conference on Acoustics, Speech and Signal Processing (ICASSP 2016), Shanghai, China, 20–25 March 2016; pp. 5200–5204. [CrossRef]
25. Papakostas, M.; Spyrou, E.; Giannakopoulos, T.; Siantikos, G.; Sgouropoulos, D.; Mylonas, P.; Makedon, F. Deep Visual Attributes vs. Hand-Crafted Audio Features on Multidomain Speech Emotion Recognition. *Computation* **2017**, *5*, 26. [CrossRef]
26. Zheng, C.; Wang, C.; Jia, N. An Ensemble Model for Multi-Level Speech Emotion Recognition. *Appl. Sci.* **2020**, *10*, 205. [CrossRef]

27. Schuller, B.; Batliner, A.; Bergler, C.; Messner, E.M.; Hamilton, A.; Amiriparian, S.; Baird, A.; Rizos, G.; Schmitt, M.; Stappen, L.; et al. The INTERSPEECH 2020 Computational Paralinguistics Challenge: Elderly Emotion, Breathing & Masks. In Proceedings of the INTERSPEECH 2020, Shanghai, China, 25–29 October 2020.
28. Kaya, H.; Karpov, A. Efficient and effective strategies for cross-corpus acoustic emotion recognition. *Neurocomputing* **2018**, *275*, 1028–1034. [CrossRef]
29. Chen, L.; Mao, X.; Wei, P.; Xue, Y.; Ishizuka, M. Mandarin emotion recognition combining acoustic and emotional point information. *Appl. Intell.* **2012**, *37*, 602–612. [CrossRef]
30. Fernandez, R.; Picard, R. Recognizing affect from speech prosody using hierarchical graphical models. *Speech Commun.* **2011**, *53*, 1088–1103. [CrossRef]
31. Nwe, T.; Foo, S.; De Silva, L. Speech emotion recognition using hidden Markov models. *Speech Commun.* **2003**, *41*, 603–623. [CrossRef]
32. Delić, V.; Bojanić, M.; Gnjatović, M.; Sečujski, M.; Jovičić, S.T. Discrimination capability of prosodic and spectral features for emotional speech recognition. *Elektron. ir Elektrotechnika* **2012**, *18*, 51–54. [CrossRef]
33. Wagner, J.; Schiller, D.; Seiderer, A.; André, E. Deep Learning in Paralinguistic Recognition Tasks: Are Hand-crafted Features Still Relevant? In Proceedings of the INTERSPEECH 2018, Hyderabad, India, 2–6 September 2018; pp. 147–151. [CrossRef]
34. Eyben, F.; Weninger, F.; Groß, F.; Schuller, B. Recent Developments in openSMILE, the Munich Open-Source Multimedia Feature Extractor. In Proceedings of the 2013 ACM Multimedia Conference, Barcelona, Spain, 21–25 October 2013; pp. 835–838. [CrossRef]
35. Akçay, M.B.; Oğuz, K. Speech emotion recognition: Emotional models, databases, features, preprocessing methods, supporting modalities, and classifiers. *Speech Commun.* **2020**, *116*, 56–76. [CrossRef]
36. Zbancioc, M.; Feraru, S. The Analysis of the FCM and WKNN Algorithms Performance for the Emotional Corpus SROL. *Adv. Electr. Comput. Eng.* **2012**, *12*, 33–38. [CrossRef]
37. Hassan, A.; Damper, R.I. Classification of emotional speech using 3DEC hierarchical classifier. *Speech Commun.* **2012**, *54*, 903–916. [CrossRef]
38. Rybka, J.; Janicki, A. Comparison of speaker dependent and speaker independent emotion recognition. *Int. J. Appl. Math. Comput. Sci.* **2013**, *23*, 797–808. [CrossRef]
39. Xu, X.; Huang, C.; Wu, C.; Wang, Q.; Zhao, L. Graph learning based speaker independent speech emotion recognition. *Adv. Electr. Comput. Eng.* **2014**, *14*, 17–22. [CrossRef]
40. Bojanić, M.; Delić, V.; Sečujski, M. Relevance of the types and the statistical properties of features in the recognition of basic emotions in the speech. *Facta Univ. Ser. Electron. Energetics* **2014**, *27*, 425–433. [CrossRef]
41. Bojanić, M.; Crnojević, V.; Delić, V. Application of neural networks in emotional speech recognition. In Proceedings of the 11th Symposium on Neural Network Applications in Electrical Engineering, Belgrade, Serbia, 20–22 September 2012; pp. 223–226. [CrossRef]
42. Bojanić, M.; Gnjatović, M.; Sečujski, M.; Delić, V. Application of dimensional emotion model in automatic emotional speech recognition. In Proceedings of the 2013 IEEE 11th International Symposium on Intelligent Systems and Informatics (SISY), Subotica, Serbia, 26–28 September 2013; pp. 353–356. [CrossRef]
43. Verkholyak, O.; Kaya, H.; Karpov, A. Modeling Short-Term and Long-Term Dependencies of the Speech Signal for Paralinguistic Emotion Classification. *SPIIRAS Proc.* **2019**, *18*, 30–56. [CrossRef]
44. Partila, P.; Tovarek, J.; Voznak, M.; Rozhon, J.; Sevcik, L.; Baran, R. Multi-Classifier Speech Emotion Recognition System. In Proceedings of the 26th Telecommunications Forum TELFOR'18, Belgrade, Serbia, 20–21 November 2018; pp. 1–4. [CrossRef]
45. Shaukat, A.; Chen, K. Emotional State Categorization from Speech: Machine vs. Human. *arXiv* **2010**, arXiv:1009.0108.
46. Holmen, K.; Ericsson, K.; Winblad, B. Quality of life among elderly: State of mood and loneliness in two selected groups. *Scand. J. Caring Sci.* **1999**, *13*, 91–95. [CrossRef]

applied sciences

MDPI

Article

EEG-Based Emotion Recognition Using Logistic Regression with Gaussian Kernel and Laplacian Prior and Investigation of Critical Frequency Bands

Chao Pan [1,2,3,*], Cheng Shi [4], Honglang Mu [3,5], Jie Li [2] and Xinbo Gao [2,6]

1 School of Computer Science and Technology, Xidian University, Xi'an 710071, China
2 School of Electronic Engineering, Xidian University, Xi'an 710071, China; leejie@mail.xidian.edu.cn (J.L.); xbgao@mail.xidian.edu.cn (X.G.)
3 College Counselor Reach Perfection with Morality Studio of Shaanxi Province, Xi'an 710071, China; hlmu@xidian.edu.cn
4 School of Computer Science and Engineering, Xi'an University of Technology, Xi'an 710048, China; chengc_s@163.com
5 Undergraduate School, Xidian University, Xi'an 710071, China
6 State Key Laboratory of Integrated Services Networks, Xi'an 710071, China
* Correspondence: cpan@xidian.edu.cn; Tel.: +86-151-2922-2993

Received: 2 February 2020; Accepted: 25 February 2020; Published: 29 February 2020

check for updates

Abstract: Emotion plays a nuclear part in human attention, decision-making, and communication. Electroencephalogram (EEG)-based emotion recognition has developed a lot due to the application of Brain-Computer Interface (BCI) and its effectiveness compared to body expressions and other physiological signals. Despite significant progress in affective computing, emotion recognition is still an unexplored problem. This paper introduced Logistic Regression (LR) with Gaussian kernel and Laplacian prior for EEG-based emotion recognition. The Gaussian kernel enhances the EEG data separability in the transformed space. The Laplacian prior promotes the sparsity of learned LR regressors to avoid over-specification. The LR regressors are optimized using the logistic regression via variable splitting and augmented Lagrangian (LORSAL) algorithm. For simplicity, the introduced method is noted as LORSAL. Experiments were conducted on the dataset for emotion analysis using EEG, physiological and video signals (DEAP). Various spectral features and features by combining electrodes (power spectral density (PSD), differential entropy (DE), differential asymmetry (DASM), rational asymmetry (RASM), and differential caudality (DCAU)) were extracted from different frequency bands (Delta, Theta, Alpha, Beta, Gamma, and Total) with EEG signals. The Naive Bayes (NB), support vector machine (SVM), linear LR with L1-regularization (LR_L1), linear LR with L2-regularization (LR_L2) were used for comparison in the binary emotion classification for valence and arousal. LORSAL obtained the best classification accuracies (77.17% and 77.03% for valence and arousal, respectively) on the DE features extracted from total frequency bands. This paper also investigates the critical frequency bands in emotion recognition. The experimental results showed the superiority of Gamma and Beta bands in classifying emotions. It was presented that DE was the most informative and DASM and DCAU had lower computational complexity with relatively ideal accuracies. An analysis of LORSAL and the recently deep learning (DL) methods is included in the discussion. Conclusions and future work are presented in the final section.

Keywords: emotion recognition; electroencephalogram (EEG); logistic regression; Gaussian kernel; Laplacian prior; affective computing

1. Introduction

Affective computing defined by Picard [1] is a multidisciplinary research field that relates to computer science, psychology, neuroscience, and cognitive science. Levenson [2] believed that during natural selection, emotions were preserved for the necessity of rapid response mechanisms when facing different environmental threats. Emotion plays a nuclear role in human behavior, such as perception, attention, decision-making, and communication [3]. Positive emotions contribute to healthy life and efficient work, while negative emotions may result in health problems [4].

Emotion recognition methods include two main categories, according to the methods humans communicate emotions, including body expressions, and physiological signals. Body expressions are physical manifestations and easy to be collected. Theorists argue that each emotion corresponds to its unique somatic response [1]. However, human physical manifestations are easily affected by the user's cultural background and social environment [4]. The physiological signals [3,4] are internal signals, such as electroencephalogram (EEG), electrocardiogram (ECG), heart rate (HR), electromyogram (EMG), and galvanic skin response (GSR). According to Connon's theory [5], the emotion changes are associated with quick responses in physiological signals coordinated by the autonomic nervous systems (ANS). This makes the physiological signals not easily controlled and overcome the shortcomings of body expressions [4]. Physiological signals have been widely applied in many studies for emotion recognition [3,4]. These physiological signals, including ECG and EMG, are still not a direct reaction to emotion changes. According to psychology and neurophysiology, emotion generation and activity have a close relationship with the activity of the cerebral cortex. Thus, EEG signals effectively reflect the brain electrical activity, and have been widely applied in many fields, including cognitive performance prediction [6], mental load analysis [7,8], mental fatigue assessment [9], recommendation system [10] and decoding visual stimuli [11,12].

Recently, the field of EEG-based emotion recognition has attracted a lot of interest, including Brain-Computer Interaction (BCI) systems, basic emotion theories, and machine learning algorithms [13,14]. In machine learning, the definition of the emotion model is necessary to describe the objective function of the algorithms. There are mainly two kinds of models [3], discrete emotion spaces and continuous emotion models. Among these models, the valence-arousal model by Russell [15] has been widely used in emotion recognition for its simplicity to establish assessment criteria. The progress of EEG-based emotion recognition also includes feature extraction, feature selection, dimension reduction, and classification algorithms [13,14]. After the pre-processing of original EEG signals, the current work is to extract and select informative features to enhance the discriminative signal characteristics. Traditionally, feature extraction and selection are based on neuroscience and cognition science [16]. For example, frontal asymmetry in Alpha band power for differentiating valence level has attracted lots of interest in neuroscience research [17]. Besides neuro-scientific assumptions, computation methods in machine learning are also applied for feature extraction and selection in EEG-based emotion recognition [16,18]. Several studies transformed the pre-processed EEG-signal into various analysis domains, including time, frequency, statistical, and spectral domains [19]. It should be noted that only one feature extraction method is not suitable for various applications and BCI systems [19]. Although the most informative EEG features for emotion classification are still being researched, the power features obtained from different bands are widely recognized as the most popular features. In these studies [20–22], power spectral density (PSD) from EEG signals worked well for identifying emotional states. However, feature extraction usually generates high-dimensional and abundant features. Feature selection and dimension reduction are necessary to avoid over-specification and to reduce computational burden [3]. Compared to filter and wrapper methods for feature selection, the dimension reduction methods, e.g., principal component analysis (PCA), and Fisher linear discriminant (FLD), are more efficient. For further information about feature selection and dimension reduction, we refer the reader to [23,24]. Many machine learning algorithms have been introduced as EEG-based emotion classifiers, such as support vector machine (SVM) [25,26], Naive Bayes (NB) [27], K-nearest neighbors (KNN), linear discriminant analysis (LDA), random forest (RF), and artificial neural networks (ANN). Among these methods, SVM based on spectral features, e.g., PSD, is the most widely applied approach. In [25], SVM was used to classify the joy, sadness, anger, and pleasure feelings based on the EEG signals from 12 symmetric electrodes pairs. SVM was used in [26] for emotion recognition with the accuracies 32% and 37% in valence and arousal dimensions, respectively. A Gaussian NB in [27] was used to classify low/high valence, and arousal emotion with precision of 57.6% and 62.0%, respectively.

Recently, deep learning (DL) methods have been introduced for EEG-based emotion classification [28,29]. The study [30,31] proposed deep belief network (DBN) to discriminate positive, neutral, and negative emotions. The experimental results show that DBN performs better than SVM and KNN. In [32], after an effective pre-processing method instead of traditional feature extraction methods, a hybrid neural network combining convolutional neural network (CNN) and recurrent neural network (RNN) is proposed to learn spatial-temporal representation from the pre-processed EEG recordings. The proposed pre-processing strategy improves the emotion recognition accuracies by about 33% and 30% for valence and arousal dimensions, respectively. In [33], a deep CNN (DCNN) model is introduced to learn discriminative representations from the combined features in the raw time domain, after normalization, and in the frequency domain. The obtained emotion classification accuracies are higher than the traditionally best bagging tree (BT) classifier. The study [34] proposed a hierarchical bidirectional gated recurrent unit (GRU) network with an attention mechanism. The proposed scheme learned more significant representation from EEG sequences and the accuracies obtained on cross-subject emotion classification task outperformed the long short time memory (LSTM) network by 4.2% and 4.6% in valence and arousal dimensions, respectively. Compared to traditional shallow methodologies, the DL models remove the signal pre-processing and feature extraction/selection progress, and are more suitable for affective representation [35,36]. However, the DL methods cannot reveal the relationship between emotional states and EEG signals for being like a black box [37]. Moreover, the training of DL networks is extremely computationally time-consuming, which limits their practical applications in real-time emotion recognition [3].

As aforementioned, the field of affective computing has developed a lot over the past several years, including the incorporation of DL methodologies. However, the modeling and recognition of emotional states is still an unexplored problem [13,14]. EEG-based emotion recognition is still faced with several challenges, including fuzzy boundaries between emotions.

Note that logistic regression (LR) [38] has been widely used as a statistical learning model in pattern recognition and machine learning, as well as in EEG signal processing. In [39], LR trained with EEG power spectral features was used for automatic epilepsy diagnosis. The work in [40] further used wavelet transform to extract effective representation from non-stationary EEG records and adopted LR as a classifier to identify epileptic and non-epileptic seizures. In [41], regularized linear LR was trained using the raw EEG signal without feature extraction to classify imaginary movements. In [42], LR with L2-penalization to avoid overfitting was trained using spectral power features from intracranial EEG (iEEG) signals for the analysis of the brain's encoding states and memory performance. The study in [43] further incorporated t-distributed stochastic neighbor embedding (tSNE) for dimension reduction of iEEG signals, and the learned L2-regularized LR classier was used for predicting memory encoding success. Despite the above studies, the potential of the LR model for EEG-based emotion recognition is still not fully explored.

In this present study, we systematically introduced the logistic regression (LR) algorithm with Gaussian kernel and Laplacian prior [44–46] for EEG-based emotion recognition. Different from these LR classifiers, Gaussian radial basis function (RBF) kernel was used to enhance the data separability in the transformed space [46]. Moreover, Laplacian prior promoting the sparsity of logistic regressors was acted as L1-regularization [44]. This prior forces many components of logistic regressors to be zero. Thus, the learned logistic regressors with sparseness control the complexity of the LR classifier and consequently avoids over-specification in EEG-based emotion recognition. The logistic regression via variable splitting and augmented Lagrangian (LORSAL) algorithm [45] was introduced to optimize the logistic regressors for lower computational complexity. Thus, the introduced LR method is abbreviated as LORSAL. For overall evaluation of the LORSAL classifier, various power spectral features and features calculated by combinations of electrodes were used as input for the classifiers. The conventional NB, SVM, linear LR with L1-regularization (LR_L1), linear LR with L2-regularization (LR_L2) were used for comparison to evaluate the performance of the LORSAL classifier. This paper also presents an investigation of critical frequency bands [47,48] and an analysis of the effect of extracted features for EEG-based emotion classification.

The rest of this paper is organized as follows. Section 2 presents the materials and methods, including the dataset for emotion analysis using EEG, physiological and video signals (DEAP), various features extracted from the EEG signals, the introduced LR model with Gaussian kernel and Laplacian prior, and the LORSAL algorithm to learn LR regressors. The experimental results are shown in Section 3. The introduced method is evaluated in the task of subject-dependent emotion recognition in valence and arousal dimensions, and the compared methods include NB, SVM, LR_L1, and LR_L2. Section 4 gives the discussion and a further comparison of LORSAL and the DL methods. Related conclusion and future work are presented in Section 5.

2. Materials and Methods

2.1. DEAP Dataset and Pre-Processing

This study was performed on the dataset DEAP developed by researchers at Queen Mary University of London [27]. This dataset is publicly available (http://www.eecs.qmul.ac.uk/mmv/datasets/deap/index.html) and consists of multimodal physiological signals for human emotion analysis. It contains, in total, 32 EEG-channel recordings and eight peripheral signals of 32 subjects (50 percent females, aged between 19 and 37). The carefully selected 40 1-min videos were used as emotion elicitation materials [27]. As shown in Figure 1, the 2D valence-arousal emotion model by Russell [15] was used to quantitatively describe emotional states. The first dimension, valence, ranges from unpleasant to pleasant, and the second dimension, arousal, changes from bored to excited. Therefore, the valence-arousal model can describe most variations in human emotion changes. The well-known self-assessment manikins (SAM) [49] (shown in Figure 2) were adopted for self-assessment along the valence and arousal dimensions, and the corresponding discrete rating values change from 1 to 9, which can be used as identification labels in emotion analysis tasks [27]. In this paper, the first 32-channel EEG records (marker in Figure 3) in the DEAP dataset preprocessed in MATLAB format were used. The EEG signals were preprocessed by down-sampling from 512 Hz to 128 Hz, and then band-pass filtering with 4–45 Hz.

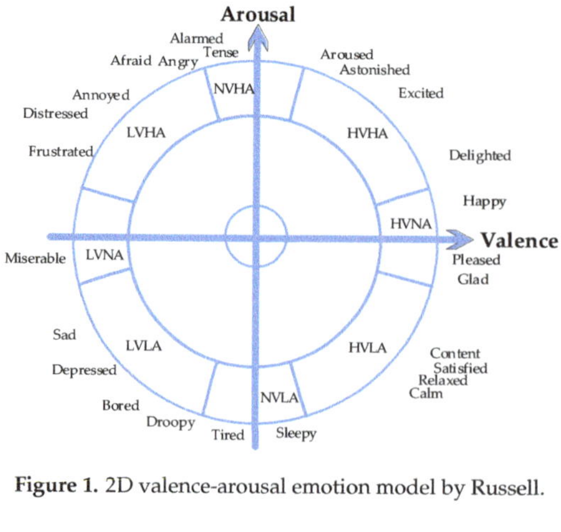

Figure 1. 2D valence-arousal emotion model by Russell.

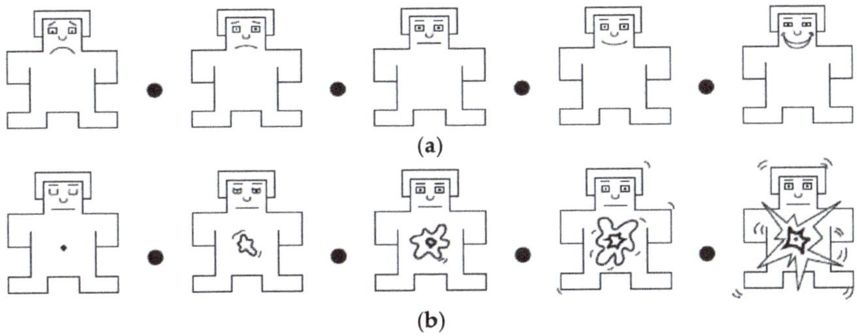

Figure 2. Images used for self-assessment manikins (SAM): (**a**) Valence SAM, (**b**) arousal SAM.

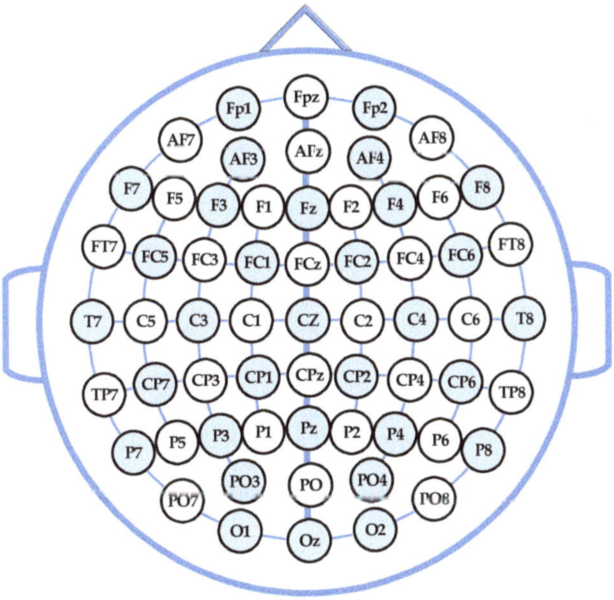

Figure 3. International 10-20 system for 32 electrodes (marked with blue circles).

In this work, two different binary classification problems were posed for subject-dependent emotion recognition: The discrimination of low/high valence (LV/HV), and low/high arousal (LA/HA). The subjects' ratings (scaling from 1 to 9) by SAM in the experiments [27] were used as the ground truth and the threshold was selected as 5 to divide the rating values into two categories: LV/HV and LA/HA. The time duration of one trail for each subject in the preprocessed EEG sequences is 63 s, in which the first 3 s are baseline signals before watching video elicitations. The 3 s sequences were removed to obtain the stimulus-related dynamics. The remaining 60-s EEG signals (thus, 7680 readings in each EEG channel, in total) were segmented into sixty 1 s epochs. Thus, there were 40 * 60 EEG epochs, in total, for each participant. Each subject-dependent EEG data had a dimensionality of 128 (sampling points) * 32 (EEG channels) * 2400 (EEG epochs). Finally, we obtained the labeled EEG signals with the dimension of 2400 for each subject. In this paper, for each subject, 10% of labeled epochs were used to train the emotion classifier, and the remaining 90% for test. For example, the constructed EEG dataset for the first participant consisted of 960 LV and 1440 HV epochs. Then, 10% of samples were randomly selected from LV and HV samples, respectively, and 240 epochs were selected for training.

Ten-fold cross-validation was used to evaluate the introduced LORSAL classifier, and the compared traditional methods.

2.2. Feature Extraction

In this study, various power spectral features in the frequency domain and features calculated by combinations of electrodes were extracted from the constructed EEG signals. The extraction of prominent statistical characteristics is important for emotion recognition. The physiological signals, e.g., EEG, are characterized with high complexity and non-stationarity, and power spectral density (PSD) [20–22] from different frequency bands is the most well-known applicable statistical feature in the task of emotion analysis. This benefits from the assumption that EEG signals are stationary for the duration of a trail [50]. Many studies in neuroscience and psychology [51] suggest that these five frequency bands are closely linked to psychological activities, including the emotion activity: Delta (1 Hz–3 Hz), Theta (4 Hz–7 Hz), Alpha (8 Hz–13 Hz), Beta (14 Hz–30 Hz), and Gamma (31–50 Hz). The fast Fourier transform (FFT) can be applied using discrete Fourier transform (DFT) [52], while the common alternatives are short-time Fourier transform (STFT) [53,54]. PSD features are extracted from the above five frequency bands using 256-point STFT and a sliding 0.5 s Hanning window with 0.25 s overlapping along 1 s epoch for each EEG channel.

Differential entropy (DE) [55,56] is a measurement of the complexity of a continuous random variable by extending the Shannon entropy concept [57]. These studies by Zheng et al. [47,48] and Duan et al. [56] introduced DE for emotion classification using EEG low/high-frequency patterns.

The original formula of DE is defined as

$$h(X) = -\int_X f(x)\log(f(x))dx, \tag{1}$$

and DE when a random variable X obeys the Gaussian distribution $N(\mu, \sigma^2)$ can be simply given as:

$$h(X) = -\int_{-\infty}^{+\infty} \frac{1}{\sqrt{2\pi\sigma^2}}\exp\frac{(x-\mu)^2}{2\sigma^2}\log\frac{1}{\sqrt{2\pi\sigma^2}}\exp\frac{(x-\mu)^2}{2\sigma^2}dx = \frac{1}{2}\log 2\pi e\sigma^2, \tag{2}$$

where π and e are constants. According to [55], given a certain frequency band, DE equals to the logarithmic spectral energy for a fixed-length EEG recording. Thus, the DE features are calculated in the five frequency bands as for the PSD features.

In the literature [58,59], the asymmetric brain activity between the left and right hemispheres is of high relation with emotions. In the studies [47,48], the differential asymmetry (DASM) and rational asymmetry (RASM) features were defined as the differences and ratios of the DE features of hemisphere asymmetric electrodes. Here, 14 pairs of asymmetric electrodes are selected to calculate DASM and RASM: Fp1-Fp2, F7-F8, F3-F4, T7-T8, P7-P8, C3-C4, P3-P4, O1-O2, AF3-AF4, FC5-FC6, FC1-FC2, CP5-CP6, CP1-CP2, and PO3-PO4. The DASM and RASM features are given as

$$\mathrm{DASM} = \mathrm{DE}(X_{left}) - \mathrm{DE}(X_{right}), \tag{3}$$

and

$$\mathrm{RASM} = \mathrm{DE}(X_{left})/\mathrm{DE}(X_{right}) \tag{4}$$

respectively. Due to the studies suggested in [58,59], the emotional states are closely linked to the spectral differences of brain activity between frontal and posterior brain regions. The definition of differential caudality (DCAU) features [47,48] was also adopted in this paper to characterize the spectral asymmetry in frontal-posterior direction. The DCAU features are given as the differences between

11 pairs of frontal-posterior electrodes: FC5-CP5, FC1-CP1, FC2-CP2, FC6-CP6, F7-P7, F3-P3, Fz-Pz, F4-P4, F8-P8, Fp1-O1, Fp2-O2. The formulation of DCAU is defined as

$$\mathrm{DCAU} = \mathrm{DE}(X_{frontal}) - \mathrm{DE}(X_{posterior}). \tag{5}$$

The dimensions of the PSD, DE, DSAM, RASM, and DCAU features are 160 (32 channels * 5 bands), 160 (32 channels * 5 bands), 70 (14 pairs of electrodes * 5 bands), 70 (14 pairs of electrodes * 5 bands), 55 (11 pairs of electrodes * 5 bands), respectively. For simplicity, the above-extracted features were used directly and separately as input for the introduced and compared recognition methods.

2.3. *Logistic Regression with Gaussian Kernel and Laplacian Prior*

Logistic regression (LR) has been a common statistical learning model in pattern recognition and machine learning [38]. Strictly speaking, the applications of LR in EEG signal analysis is not new, as illustrated in the above introduction [39–43]. Despite this, the potential of the LR model for EEG-based emotion recognition has not been fully exploited. In this paper, we systematically introduced the logistic regression (LR) algorithm with Gaussian kernel and Laplacian prior [44–46] for emotion recognition with EEG signals.

The goal of a supervised learning algorithm is to train a classifier using training samples in order to recognize the label of an input feature vector from different classes. In EEG-based emotion recognition, the major task is to assign the input EEG signals to one of the given classes. Especially in this study, two binary classification problems were posed for subject-dependent emotion recognition: The classification of LV/HV emotions, and LA/HA emotions.

Using a multinomial LR (MLR) model [38,44], the probability that the input feature $\mathbf{x}_i$ belongs to emotion class k is written as

$$p(y_i = k|\mathbf{x}_i, w) = \frac{\exp(w^{(k)}\mathbf{h}(\mathbf{x}_i))}{\sum_{k=1}^{K}\exp(w^{(k)}\mathbf{h}(\mathbf{x}_i))}, \tag{6}$$

where $\mathbf{x}_i$ is the feature vector extracted from the original EEG sequences, and $\mathbf{h}(\mathbf{x}_i)$ indicates a vector of functions of the input feature vector $\mathbf{x}_i$, and $w \equiv [w^{(1)^{\mathrm{T}}}, \ldots, w^{(K)^{\mathrm{T}}}]^{\mathrm{T}}$ is the logistic regressors. For binary classification tasks ($K = 2$), this is known as LR model, for $K > 2$, the usual designation is MLR [44]. Although emotion recognition in this paper is binary classification, the formula of MLR is presented here for completeness. On one hand, this does not affect the understanding of the model, on the other hand, this makes it easy to extend to the cases when handling multiple emotion classes.

Note that the function $\mathbf{h}(\mathbf{x}_i)$ can be linear or nonlinear. For the latter case, kernel functions can be selected to further enhance the separability of extracted features in the transformed space. In this study, the Gaussian kernel is utilized, given by

$$K(\mathbf{x}_i, \mathbf{x}_j) = \exp(-\parallel \mathbf{x}_i - \mathbf{x}_j \parallel / (2\rho^2)), \tag{7}$$

In this paper, the training of the LR classifier using labeled EEG epochs amounts to estimate the class densities and learn the logistic regressor w. Following the formulation of the sparse MLR (SMLR) algorithm in [44], the solution of w is given by the maximum a posteriori (MAP) estimate

$$\hat{w} = \underset{w}{\operatorname{argmax}}\,\ell(w) + \log p(w), \tag{8}$$

where $\ell(w)$ indicates the log-likelihood function given as following:

$$\ell(w) = \log \prod_{i=1}^{L} p(y_i|\mathbf{x}_i, w), \tag{9}$$

where L denotes the number of training samples, and

$$p(\boldsymbol{w}) \propto \exp(-\lambda \parallel \boldsymbol{w} \parallel_1), \tag{10}$$

denotes the Laplacian prior, where $\parallel \boldsymbol{w} \parallel_1$ indicates the L1 norm of $\boldsymbol{w}$, and l is the regularization parameter. The Laplacian prior forces the sparsity on the logistic regressors $\boldsymbol{w}$, and promote many components of $\boldsymbol{w}$ equal to zero [45,46]. The obtained sparse regressor reduces the complexity of the LR classifier and, therefore, avoids over-specification in EEG-based emotion classification.

The convex problem in Equation (8) is difficult to optimize for the nonquadratic property of the term $\ell(\boldsymbol{w})$ and the non-smoothness of the term $\log p(\boldsymbol{w})$. The studies in [44,60] decomposed the problem in Equation (8) into a sequence of quadratic problems using a majorization-minimization scheme [61]. The SMLR algorithm optimizes each quadratic problem with the complexity of $O(((L+1)K)^3)$ [44]. The fast SMLR (FSMLR) [62] is more efficient by applying a block-based Gauss–Seidel iterative procedure to estimate $\boldsymbol{w}$. Thus, the FSMLR algorithm is K^2 faster than SMLR with the complexity of $O((L+1)^3K)$.

In this work, the logistic regression via variable splitting and augmented Lagrangian (LORSAL) [45] algorithm is introduced to solve the LR regressors in Equation (8). LORSAL has been proposed for hyperspectral image classification (HSI) in remote sensing community [45,46]. The complexity of LORSAL is $O((L+1)^2K)$ for each quadratic problem, compared to the $O(((L+1)K)^3)$ and $O((L+1)^3K)$ complexities of the SMLR and FSMLR algorithms. Note that in this paper, we might use LORSAL directly to indicate the introduced LR with Gaussian kernel and Laplacian prior.

3. Experimental Results

In this work, we systematically investigated the classification performance of the introduced LORSAL method compared with four classifiers, Naive Bayes (NB) [27], support vector machine (SVM) [25,26], linear LR with L1-regularization (LR_L1), linear LR with L2-regularization (LR_L2) for the binary classification of the LV/HV and LA/HA emotional states. These features, including PSD, DE, DASM, RASM, and DCAU, were extracted from the EEG-signals and used directly as inputs for the classifier. The NB in MATLAB was employed as in [27]. The LIBLINEAR [63] software was adopted for the implementation of the LR_L1, and LR_L2 classifiers, respectively, with the default cost parameter. The LIBSVM [64] tool was utilized to implement the SVM classifier by using the linear kernel with default parameters. For simplicity, the parameters for Gaussian kernel and Laplacian prior in the LORSAL method were set as default in [46]. Such parameter settings may be not optimal for EEG-based emotion recognition, but present ideal classification performance in the experiments.

3.1. Overall Classification Accuracy

The mean accuracies and standard deviations obtained by different classifiers in valence dimension for different features extracted from five frequency bands (Delta, Theta, Alpha, Beta, and Gamma) and the total frequency bands are tabulated in Table 1. It should be noted that 'Total' in Table 1 denotes the features by concatenating all different features from all frequency bands. Given the same features extracted from the EEG signals, the accuracy metrics of the black bold font in Table 1 indicate the highest accuracies obtained by different classifiers for each frequency band, while the precision metrics with gray background denote the highest precisions obtained by compared methods for all kinds of frequency bands. The LORSAL methods obtained the highest accuracy of 77.17% for the DE feature from the total frequency bands among all the compared classifiers. Under the same case, the highest classification accuracy obtained by SVM is 69.55%, while the best accuracy by NB is 62.36% for the DASM feature from the total frequency bands.

Table 1. The mean precisions and standard deviations (%) of the classification of LV/HV emotions obtained by the compared classifiers for different features extracted from different frequency bands.

Feature	Classifier	Delta	Theta	Alpha	Beta	Gamma	Total
PSD	NB	54.18/5.39	53.94/5.24	54.25/5.08	58.85/7.63	61.23/8.48	60.51/6.71
	SVM	51.14/14.31	47.57/15.75	52.35/16.04	62.97/12.18	64.15/12.19	**69.04/5.91**
	LR_L1	44.80/4.63	45.23/4.72	44.00/5.51	38.41/8.75	36.18/9.67	34.43/9.52
	LR_L2	44.57/4.88	45.11/4.92	44.14/5.49	38.47/8.63	36.38/9.58	34.38/9.57
	LORSAL	**56.45/5.79**	**56.19/5.34**	**58.29/5.56**	**64.95/6.78**	**68.30/7.98**	63.29/6.54
DE	NB	54.27/3.62	53.90/3.53	54.62/4.07	59.19/5.59	61.52/6.51	61.04/5.76
	SVM	52.08/13.54	49.76/14.14	52.04/13.41	62.91/10.23	65.88/9.93	69.55/6.58
	LR_L1	44.04/4.80	44.67/4.73	43.16/5.34	38.22/8.67	36.08/9.52	33.69/9.88
	LR_L2	43.86/4.88	44.44/4.69	43.15/5.41	38.15/8.64	36.06/9.52	33.81/9.59
	LORSAL	**61.79/4.55**	**58.34/3.90**	**59.20/4.04**	**67.06/6.95**	**72.93/7.30**	**77.17/6.37**
DASM	NB	54.81/6.61	53.97/5.53	54.07/6.12	60.18/6.16	62.27/6.88	62.36/5.66
	SVM	45.64/14.11	44.66/14.77	44.83/14.66	56.24/13.18	60.95/12.14	64.48/10.21
	LR_L1	45.32/4.11	45.77/3.89	45.41/4.86	40.68/7.45	38.83/8.79	36.90/8.82
	LR_L2	45.35/4.04	46.03/3.95	45.59/4.66	40.77/7.53	38.84/8.68	37.11/8.80
	LORSAL	**58.38/3.58**	**55.21/3.55**	**55.42/3.46**	**60.31/5.39**	**64.63/6.22**	**71.63/4.94**
RASM	NB	**51.17/6.55**	51.61/7.37	51.02/6.98	**54.90/8.57**	**58.08/9.47**	**55.65/6.77**
	SVM	37.41/15.45	37.23/14.33	37.35/13.87	40.56/16.24	47.29/17.02	48.17/17.72
	LR_L1	49.39/6.07	48.19/5.55	48.17/6.24	45.29/6.11	42.39/7.44	42.71/6.64
	LR_L2	49.45/4.75	47.92/4.30	47.95/4.73	45.44/5.87	42.53/6.97	42.69/6.37
	LORSAL	49.31/8.79	**51.68/8.71**	**51.68/8.98**	52.55/10.36	57.38/9.69	51.67/7.53
DCAU	NB	53.98/6.22	52.97/6.73	53.72/6.03	**59.61/5.70**	61.97/6.52	61.95/5.52
	SVM	43.40/13.46	41.11/13.93	43.91/14.39	55.47/13.83	59.57/12.67	63.48/9.93
	LR_L1	45.94/4.24	46.26/4.11	45.82/4.41	41.38/7.32	39.19/8.29	37.41/8.21
	LR_L2	46.03/4.23	46.45/4.11	45.74/4.38	41.33/7.24	39.21/8.35	37.45/8.22
	LORSAL	**57.01/3.38**	**54.68/3.59**	**55.20/3.55**	59.38/5.61	**63.54/6.21**	**69.89/4.89**

The SVM classifier is the most widely applied approach based on spectral features, especially PSD. Table 1 shows that the performance of SVM is second only to LORSAL on all features extracted from total frequency bands, and the corresponding accuracies are 69.04%, 69.55%, 64.48%, 48.17%, and 63.48% for the PSD, DE, DASM, RASM, and DCAU features. In study [27], the NB classifier obtained an accuracy of 57.6% in the valence dimension. In this study, the mean precisions obtained by NB are approximately between 60% and 62% for the PSD, DE, DASM, and DCAU features from the total frequency bands.

However, the best accuracies obtained by LR_L1 and LR_L2 are approximately 46%, which is significantly lower than those obtained by NB, SVM, and LORSAL. Although the LR_L1 and LR_L2 adopted L1-regularization and L2-regularization during the optimization of LR regressors to avoid over-specification, the assumption of linear separability does not hold for the extracted features from EEG signals. The average precisions obtained by the LORSAL have significant improvement over these by LR_L1 after incorporating the Gaussian kernel. The Gaussian kernel can enhance the data separability in the transformed space and meanwhile, the Laplacian prior can promote sparsity on the learned LR regressor and avoid over-specification of the selected training EEG epochs.

For the classification task of LV/HV emotions, the LORSAL methods present the best classification accuracies, 77.17%, 71.63%, and 69.89% on the DE, DASM, and DCAU features from total frequency bands, which are higher than SVM by about 8%, 7%, and 6%. The SVM and NB classifiers perform best by accuracies 69.04% and 55.65% on the PSD and RASM features from 'Total' bands, respectively. As is shown in Table 2, the performance of the five classifiers on classifying LA/HA emotions is similar to the case of LV/HV classification. The introduced LORSAL method performed best on the DE, DASM, and DCAU features from 'Total' bands, and the corresponding accuracies outperformed these of SVM by about 7%, 9%, and 8%. In addition, the performance of LORSAL was obviously better than that of the compared LR_L1, and LR_L2 methods in the arousal dimension. The incorporated Gaussian kernel and Laplacian prior improved the distinguishing ability of LORSAL in emotion recognition task based on EEG signals.

Table 2. The mean precisions and standard deviations (%) of the classification of LA/HA emotions obtained by the compared classifiers for different features extracted from different frequency bands.

Feature	Classifier	Delta	Theta	Alpha	Beta	Gamma	Total
PSD	NB	54.22/5.37	53.08/5.66	53.82/5.30	55.39/8.03	58.44/8.94	57.54/6.15
	SVM	50.83/15.93	48.07/15.6	49.95/15.65	58.05/15.31	58.82/15.08	**68.60/8.07**
	LR_L1	47.74/6.07	47.96/5.66	47.46/6.49	45.34/9.97	44.11/11.67	43.55/14.04
	LR_L2	47.66/6.31	47.94/5.78	47.42/6.41	45.47/9.96	44.03/11.64	43.58/14.23
	LORSAL	**57.45/6.75**	**56.91/6.86**	**58.82/6.23**	**63.99/6.47**	**67.75/7.05**	61.62/6.64
DE	NB	54.13/3.76	53.43/3.68	54.08/3.76	56.97/4.81	58.75/5.33	58.46/5.00
	SVM	46.90/14.53	45.03/14.93	47.51/14.57	55.88/13.34	63.33/12.79	69.92/7.94
	LR_L1	47.29/6.52	47.69/5.93	46.96/6.81	45.26/10.88	44.16/12.99	43.10/15.21
	LR_L2	47.38/6.40	47.81/5.89	46.91/6.88	45.29/11.03	44.12/13.03	43.25/15.00
	LORSAL	**61.97/4.64**	**58.18/3.94**	**59.35/3.97**	**66.57/6.67**	**72.73/7.62**	**77.03/6.20**
DASM	NB	54.67/6.10	54.78/6.33	54.31/7.68	58.11/5.75	60.46/5.28	60.34/4.50
	SVM	43.95/13.24	42.84/13.38	43.02/12.53	51.24/14.69	56.42/13.68	61.79/12.34
	LR_L1	48.07/5.75	48.16/4.98	47.91/5.34	45.99/8.41	45.45/10.67	44.60/12.33
	LR_L2	48.13/5.64	48.05/5.18	47.76/5.38	45.93/8.45	45.43/10.72	44.78/12.22
	LORSAL	**58.13/3.75**	**55.14/3.62**	**55.18/3.37**	**59.70/4.94**	**64.54/5.87**	**71.20/4.96**
RASM	NB	**51.31/7.57**	**51.92/7.12**	**51.01/6.47**	**54.25/6.52**	**55.26/6.84**	**53.57/4.25**
	SVM	36.59/13.04	35.31/11.31	36.18/12.49	37.79/14.05	42.61/17.12	43.46/15.61
	LR_L1	49.45/6.07	49.07/4.97	50.18/4.32	49.14/5.75	47.83/7.35	47.94/7.39
	LR_L2	49.44/5.08	49.16/4.41	50.19/4.22	49.05/5.73	47.92/7.21	48.10/7.31
	LORSAL	49.09/10.60	50.93/10.79	50.15/10.95	50.14/11.36	53.51/11.07	50.67/8.33
DCAU	NB	55.09/7.10	53.86/6.91	53.98/7.42	56.98/7.23	60.14/5.99	59.98/4.50
	SVM	42.34/13.47	41.56/13.26	43.19/13.26	50.64/13.76	55.06/14.99	60.04/12.79
	LR_L1	48.19/5.42	48.36/4.90	48.01/5.18	46.34/8.24	45.29/10.05	44.47/11.45
	LR_L2	48.21/5.31	48.38/5.04	47.86/5.14	46.29/8.13	45.33/9.93	44.61/11.51
	LORSAL	**57.16/3.69**	**54.49/3.65**	**55.18/3.46**	**58.09/4.99**	**62.68/5.60**	**68.48/4.93**

For a more comprehensive comparison of the NB, SVM, and LORSAL approaches, Table 3 tabulated the average values and standard deviations of precision, recall, and F1, for the binary emotion classification problem of LV/HV and LA/HA, respectively, when different features were extracted from the total frequency bands. The introduced LORSAL method obtained the best precisions (77.17%/6.37% for LV/HV, and 77.03%/6.20% for LA/HA), the best recalls (76.79%/6.21% for LV/HV, and 76.15%/6.14% for LA/HA), and the best F1 values (76.90%/6.27% for LV/HV, and 76.47%/6.14% for LA/HA), for EEG-based emotion recognition. In summary, the above analysis suggests the application of LORSAL on the DE features extracted from 'Total' bands for EEG-based emotion recognition. For simplicity, we will focus on comparing the performance of LORSAL with NB and SVM in the following subsections.

Table 3. The mean metrics and standard deviations (%) of precision, recall, and F1 for the binary classification of LV/HV and LA/HA emotions obtained by the compared classifiers for different features extracted the total frequency bands.

Feature	Classifier	Valence			Arousal		
		Precision	Recall	F1	Precision	Recall	F1
PSD	NB	60.51/6.71	56.65/4.99	48.84/9.54	57.54/6.15	54.95/4.29	46.86/8.54
	SVM	69.04/5.91	65.62/6.55	65.24/7.39	68.60/8.07	61.09/6.17	60.41/7.47
	LORSAL	63.29/6.54	62.15/6.65	61.84/7.27	61.62/6.64	59.02/5.53	58.46/6.54
DE	NB	61.04/5.76	60.31/5.19	58.83/5.48	58.46/5.00	58.96/5.05	55.84/5.74
	SVM	69.55/6.58	66.93/6.50	66.89/7.15	69.92/7.94	63.50/6.57	63.45/7.54
	LORSAL	77.17/6.37	76.79/6.21	76.90/6.27	77.03/6.20	76.15/6.14	76.47/6.14
DASM	NB	62.36/5.66	62.18/5.54	61.73/5.57	60.34/4.50	60.79/4.90	59.76/4.77
	SVM	64.48/10.21	63.01/7.15	62.01/9.11	61.79/12.34	59.06/6.36	57.49/8.46
	LORSAL	71.63/4.94	71.38/4.92	71.43/4.92	71.20/4.96	70.68/5.00	70.82/4.94
RASM	NB	55.65/6.77	53.72/4.64	46.01/7.92	53.57/4.25	52.38/2.95	40.70/6.37
	SVM	48.17/17.72	52.47/5.24	43.24/9.56	43.46/15.61	50.25/0.98	40.00/3.72
	LORSAL	51.67/7.53	51.35/2.98	46.69/5.49	50.67/8.33	50.60/2.11	45.32/3.94
DCAU	NB	61.95/5.52	61.65/5.44	61.26/5.48	59.98/4.50	60.14/4.50	59.36/4.37
	SVM	63.48/9.93	62.09/6.51	60.99/8.59	60.04/12.79	58.21/6.32	56.12/8.76
	LORSAL	69.89/4.89	69.68/4.77	69.71/4.81	68.48/4.93	68.11/4.90	68.19/4.88

3.2. Investigation of Critical Frequency Bands

In this study, the informative features were extracted from different frequency bands (Delta, Theta, Alpha, Beta, Gamma, and Total) for EEG-based emotion recognition. Thus, we present an investigation of the critical frequency bands in EEG signals for emotion processing. Figures 4a–f and 5a–f show the mean precisions obtained by LORSAL, SVM, and NB for the classification of LV/HV and LA/HA, respectively, when the frequency bands alternated among Delta, Theta, Alpha, Beta, Gamma, and Total. Gamma and Beta are more informative than other frequency bands (Delta, Theta, and Alpha). For example, among the first five frequency bands, the LOSAL obtained the highest accuracies of 72.93% and 67.06% on Gamma and Beta from the DE features in valence classification, the best accuracies of 72.73% and 66.57% on Gamma and Beta from the DE features in arousal recognition.

There is not always a causal relationship between features with high recognition accuracies and emotions. Koelstra et al. [27] presented an investigation about the causal relationship between the emotions and their EEG signals on the DEAP dataset. The average frequency power of trials was calculated over the bands Theta, Alpha, Beta, and Gamma (between 3 and 47 Hz). The Spearman correlated coefficients were tabulated [27] to present the statistical correlation of the power changes of EEG sequences and subject ratings. Similar research has been done by Zheng [47], we focused on analyzing the informative neural patterns associated with recognizing different emotions. Especially, the Fisher ratio was used to investigate the critical frequency bands for discriminating different emotions. The Fisher ratio has been used in pattern recognition for class separability measure and feature selection [65–67], as well as in emotion classification [3,4,13,14,27]. The higher values of the Fisher ratio mean more informative neural patterns and features related to emotion recognition. It is defined as the ratio of interclass difference to the intraclass spread

$$F_n(\mathrm{L},\mathrm{H}) = \frac{(\mathrm{m}_{Ln} - \mathrm{m}_{Hn})^2}{\sigma_{Ln}^2 + \sigma_{Hn}^2}, \quad (11)$$

where L and H mean two different emotion, e.g., LV/HV or LA/HA, and m_{Ln}, m_{Hn}, σ_{Ln}^2, and σ_{Hn}^2 denote the mean and variance of the n-th dimension of the EEG feature belonging to emotions L and H, respectively. Thus, $F_n(\text{L}, \text{H})$ indicates the class separability between emotions L and H for the n-th dimension of the extracted feature.

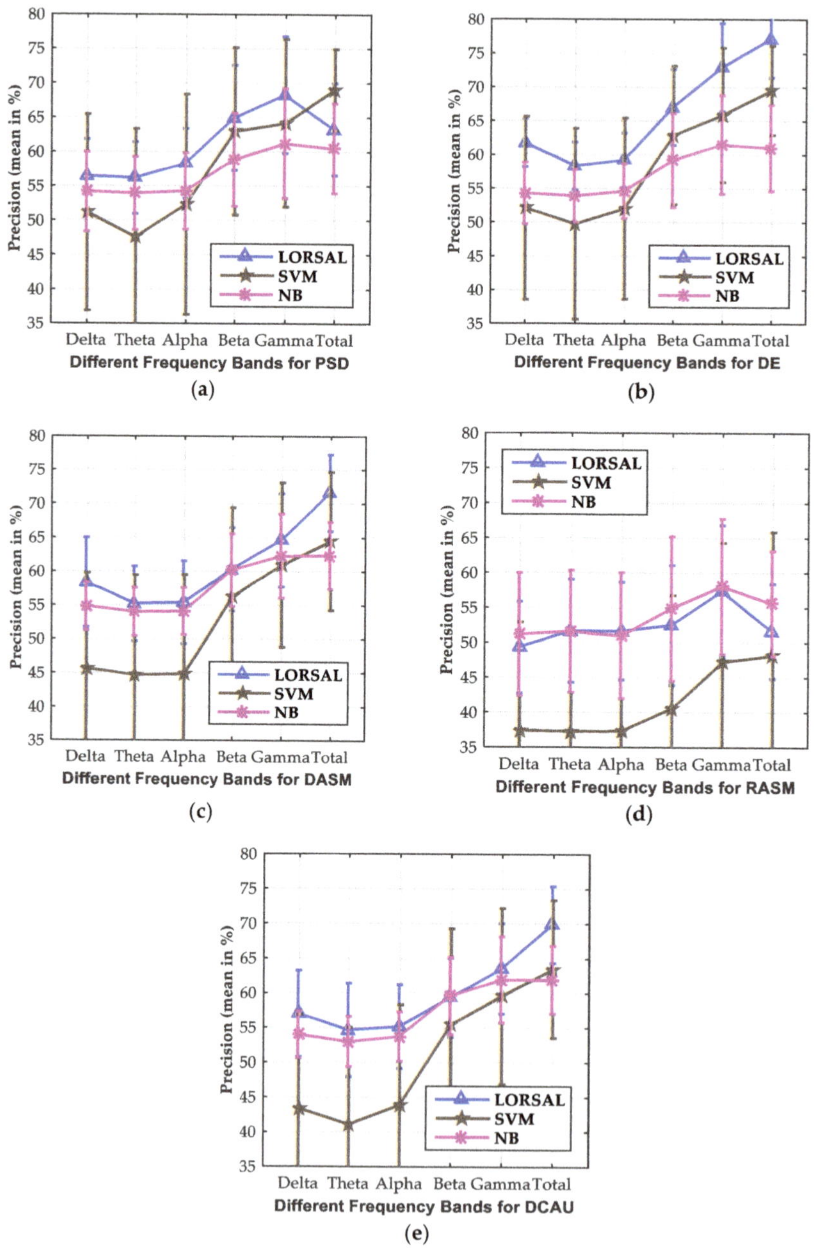

Figure 4. Effects of the different frequency bands (Delta, Theta, Alpha, Beta, Gamma, and Total) on the classification precisions for LV/HV emotions obtained by the LORSAL, SVM, and NB classifiers for the five different features: **(a)** PSD, **(b)** DE, **(c)** DASM, **(d)** RASM, **(e)** DCAU.

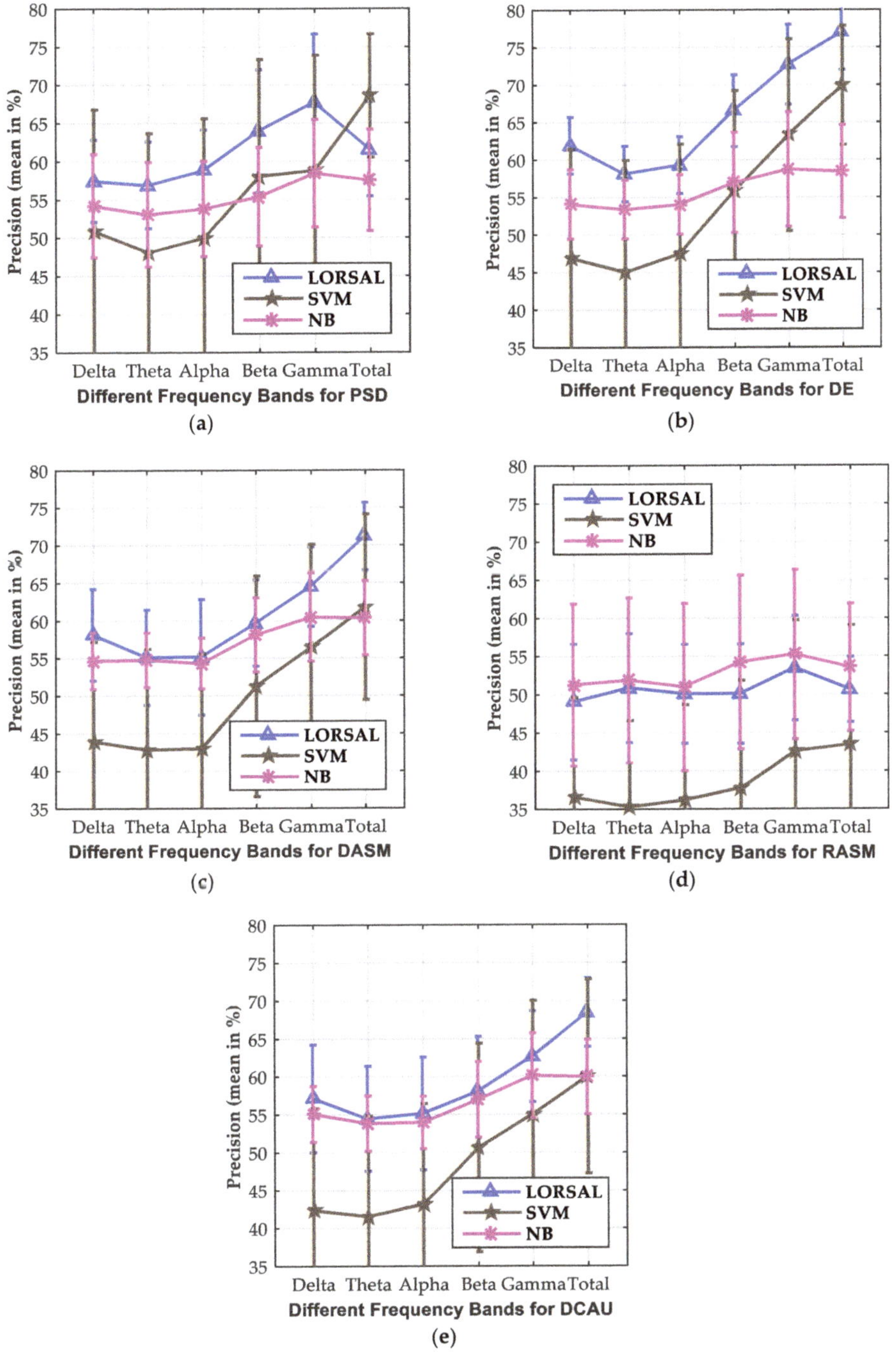

Figure 5. Effects of the different frequency bands (Delta, Theta, Alpha, Beta, Gamma, and Total) on the classification precisions for LA/HA emotions obtained by the LORSAL, SVM, and NB classifiers for the five different features: (**a**) PSD, (**b**) DE, (**c**) DASM, (**d**) RASM, (**e**) DCAU.

Given the extracted feature and the specific frequency bands, the mean Fisher ratio $F(\mathrm{L},\mathrm{H})$ is calculated by averaging the values of $F_n(\mathrm{L},\mathrm{H})$ over all the EEG channels (e.g., PSD and DE) or combinations of electrodes (e.g., DASM, RASM, and DCAU)

$$F(\mathrm{L},\mathrm{H}) = \frac{1}{N}\sum_{n=1}^{N} F_n(\mathrm{L},\mathrm{H}), \tag{12}$$

where N is the number of EEG channels or combinations of electrodes.

Figure 6 shows the Fisher ratio of the extracted PSD, DE, DASM, RASM, and DCAU feature along different frequency bands by averaging the values over all subjects in valence and arousal dimensions, respectively. In addition, Table 4 illustrates the Fisher ratio over different frequency bands by averaging the values over all features and subjects in valence and arousal dimensions, respectively. The Fisher ratio values shown in Table 4 are calculated by further averaging the values presented in Figure 6 over the five different frequency bands. The following subsection presents comprehensively an analysis including the EEG neural patterns associated with emotions in previous studies [27,47,48], the critical frequency bands and the informative features for emotion recognition. Specific frequency ranges are highly related to certain brain activities. There are neuroscience findings [68,69] revealing that the Alpha bands in EEG signals associate with attentional processing, while the Beta bands associate with emotional and cognitive progress. In the previous studies on the DEAP dataset by Koelstra et al. [27], they reported negative correlations in the Theta, Alpha, and Gamma bands for arousal, and strong correlations in all investigated frequency bands for valence. Similarly, Onton et al. [70] found a positive correlation of valence and the Beta and Gamma bands.

From Figure 6 and Table 4, we found that the Gamma and Beta bands obtained higher values of the Fisher ratio than the other frequency bands. This means that the features extracted over Gamma and Beta bands are more effective for discriminating different emotions, as is in accordance with the classification accuracies by the compared classifiers illustrated in Figure 5. Similarly, Li and Lu [71] showed that the EEG Gamma bands are appropriate for emotion recognition when images are used for emotion elicitation. The studies by [47] Zheng and Lu found specific neural patterns in high-frequency bands for distinguishing negative, neutral, and positive emotions. For negative and neutral emotions, the energy of Beta and Gamma frequency bands decreases, while positive emotions present higher energy of these two frequency bands. Their experimental results [47] on SJTU Emotion EEG Dataset (SEED) showed that the KNN, LR_L2, SVM, and DBN classifiers performed better on Gamma and Beta frequency bands than other bands for the PSD, DE, DASM, RASM, and DCAU features. This showed the informativeness of EEG Gamma and Beta bands for emotion recognition with film clips as stimuli in the SEED dataset. The emotion elicitation materials in the DEAP dataset are one-minute videos [27]. Our experimental results in DEAP and findings were in accordance with these previous studies [47,48]. Additionally, the total frequency bands concatenating all the original five frequency bands can further improve the emotion recognition performance, and the LORSAL obtained the highest mean accuracies in valence and arousal dimensions from the DE, DASM, and DCAU features, as is consistent with the results in studies [47,48].

Table 4. Fisher ratio of different frequency bands by averaging the values over all features and subjects in valence and arousal dimensions, respectively.

Emotion	Valence					Arousal				
Frequency Bands	Delta	Theta	Alpha	Beta	Gamma	Delta	Theta	Alpha	Beta	Gamma
PSD	0.046	0.005	0.014	0.155	**0.414**	0.045	0.005	0.013	0.124	**0.418**
DE	0.047	0.052	0.077	0.242	**0.256**	0.051	0.048	0.062	0.173	**0.197**
DASM	0.057	0.063	0.070	0.240	**0.317**	0.068	0.073	0.067	0.186	**0.243**
RASM	0.004	0.050	0.037	**0.131**	0.115	0.004	0.065	0.052	**0.095**	0.080
DCAU	0.059	0.062	0.064	0.224	**0.307**	0.062	0.064	0.067	0.195	**0.244**

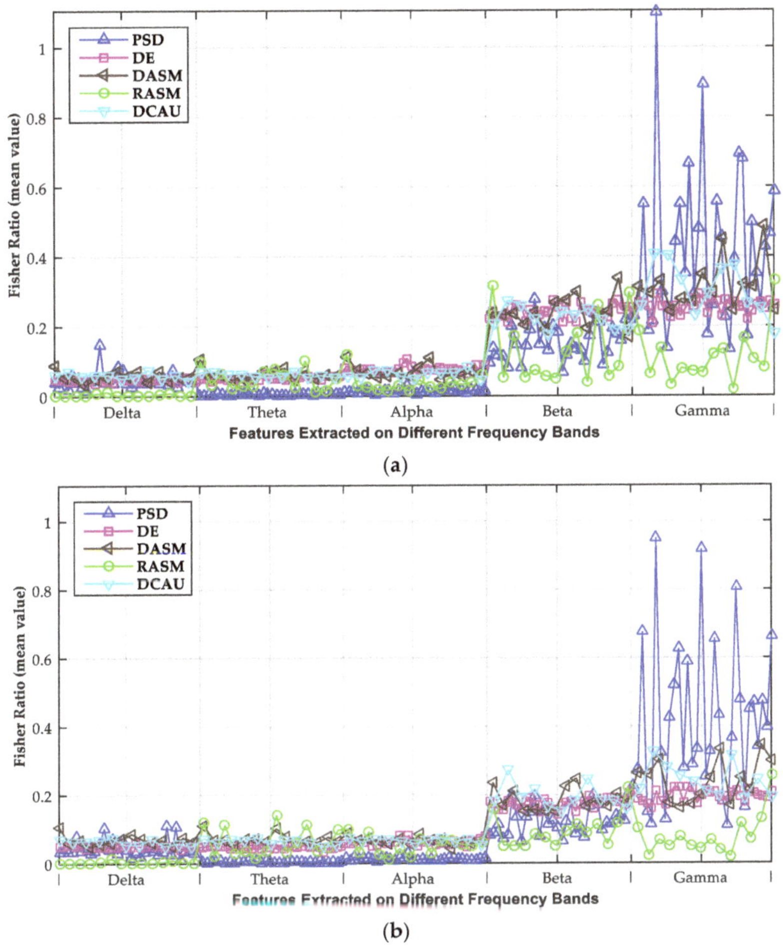

Figure 6. Fisher ratio of features extracted on different frequency bands by averaging the values over all subjects in **(a)** valence and **(b)** arousal dimensions, respectively.

3.3. *Effect of Extracted Features*

This subsection presents an analysis of the effects of different features on the average accuracies for emotion recognition based on the EEG signal. When the extracted features alternate among PSD, DE, DASM, RASM, and DCAU, the mean precisions obtained by LORSAL, SVM, and NB for the classification of LV/HV and LA/HA are shown in Figures 7a–f and 8a–f. All the LORSAL, SVM, and NB classifiers performed best for the DE features. Among all classifiers, LORSAL obtained the highest accuracies of 77.17% and 77.03% in valence and arousal dimensions, respectively, for the DE features extracted from total frequency bands. In addition, the highest precisions obtained by SVM are 69.55% and 69.92%, respectively, for DE from total frequency bands. The DE features denote the complexity of continuous random variables [55–57]. The EEG signals are characterized by higher low-frequency energy over high-frequency energy, and consequently, DE can distinguish EEG sequences according to low- and high-frequency energy. These results agree with the findings in [47,48], and further show the superiority of the DE features in EEG-based emotion classification.

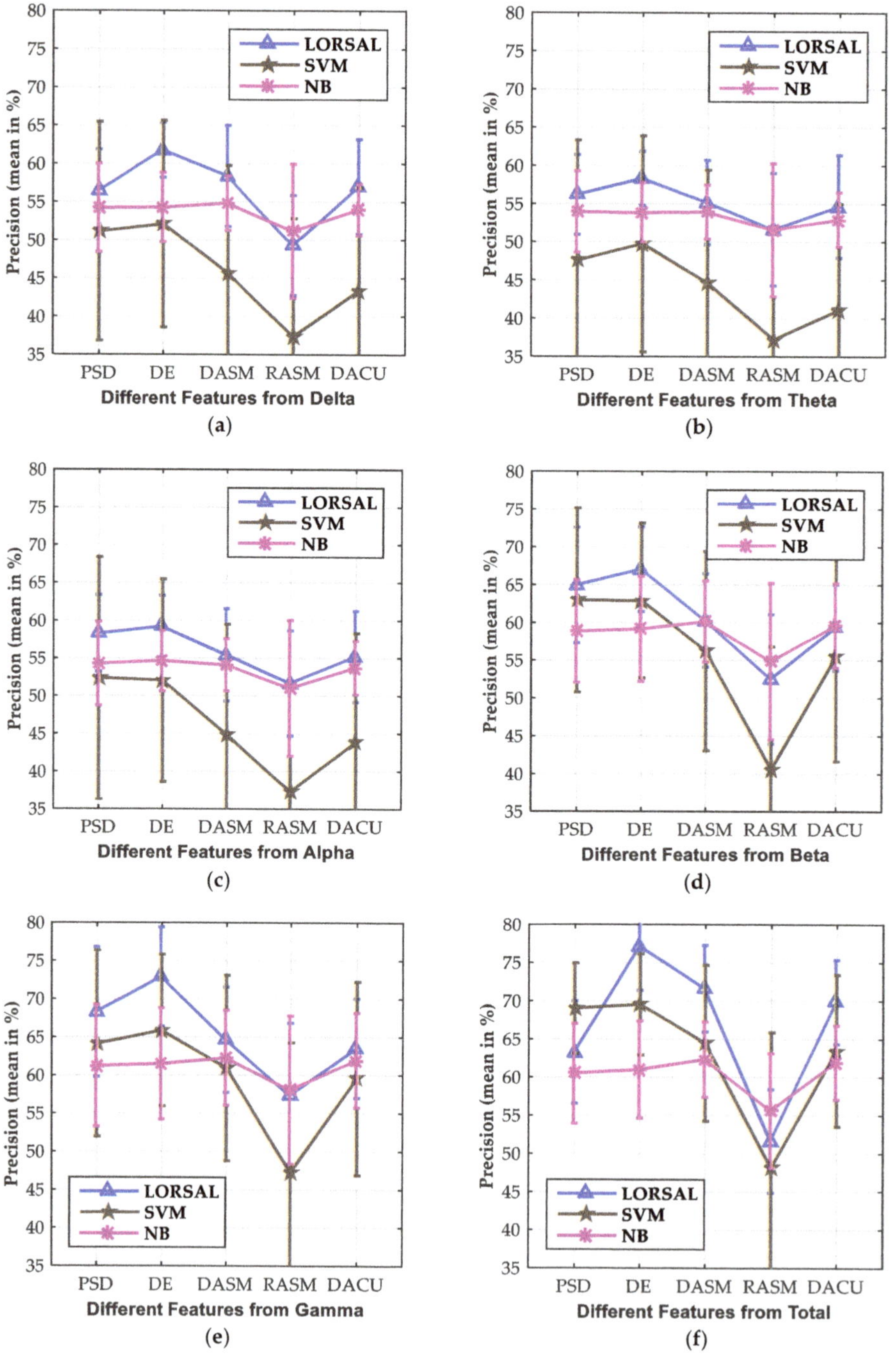

Figure 7. Effects of the different features (PSD, DE, DASM, RASM, and DCAU) on the classification precisions for LV/HV emotions obtained by the LORSAL, SVM, and NB classifiers from the six different frequency bands: (**a**) Delta, (**b**) Theta, (**c**) Alpha, (**d**) Beta, (**e**) Gamma, (**f**) Total.

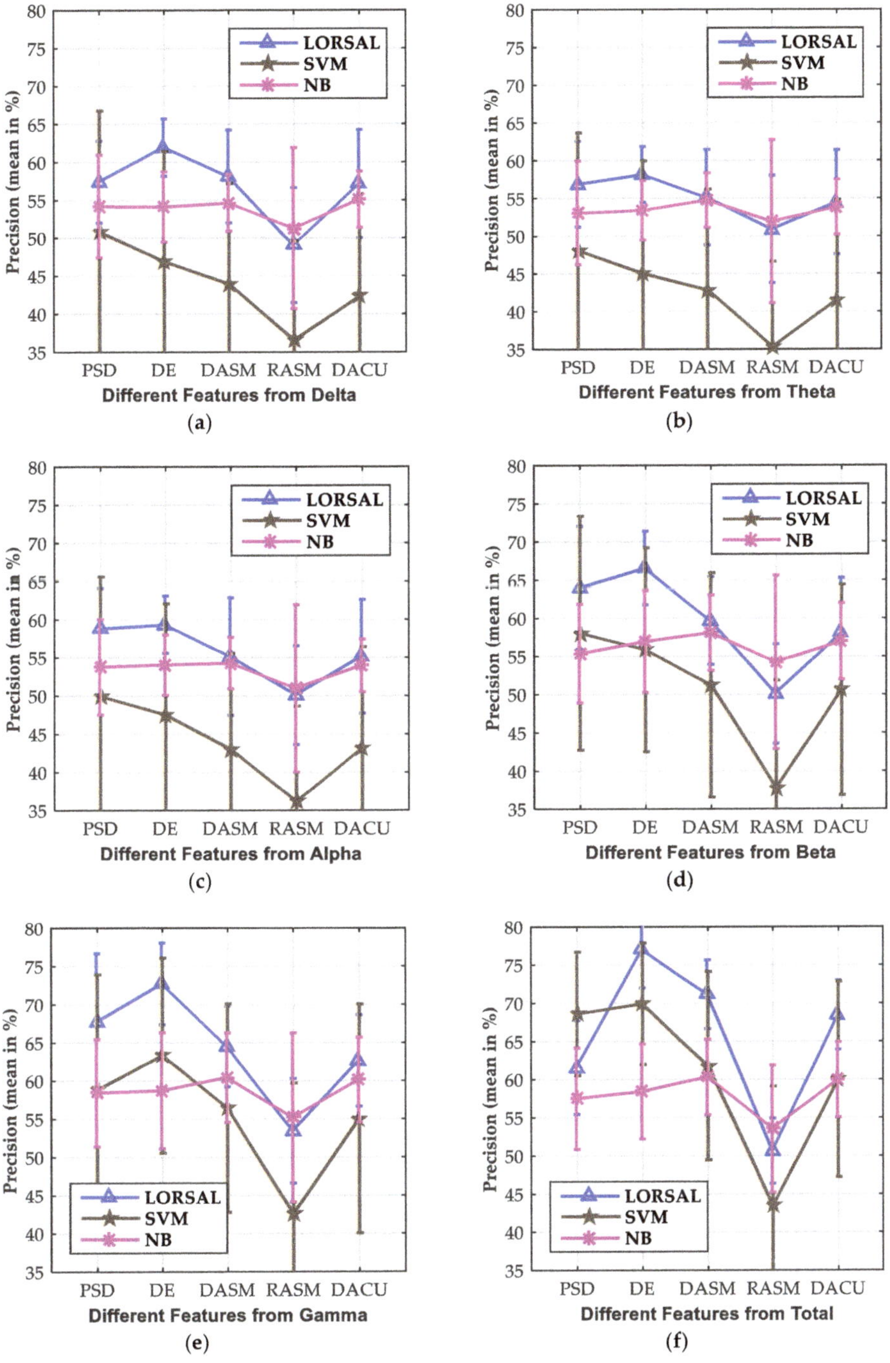

Figure 8. Effects of the different features (PSD, DE, DASM, RASM, and DCAU) on the classification precisions for LA/HA emotions obtained by the LORSAL, SVM, and NB classifiers from the six different frequency bands: (**a**) Delta, (**b**) Theta, (**c**) Alpha, (**d**) Beta, (**e**) Gamma, (**f**) Total.

Moreover, the DASM and DCAU features provide relatively ideal performance compared to the PSD and DE features. DASM and DCAU are asymmetric features and the former findings showed the effectiveness of asymmetrical brain activity along left-right and frontal-posterior directions in emotion analysis. It is noted that the dimensions of the DASM and DCAU features are 70 and 55, respectively, which are fewer than that of PSD and DE with 160-dimension features. This makes DASM and DCAU more competitive in computational complexity. The experimental results were also consistent with the findings by Zheng and Lu [47,48].

4. Discussion

Although the area of affective computing has developed a lot over the past years, the topic of EEG-based emotion recognition is still a challenging problem. This paper introduced LR with Gaussian kernel and Laplacian prior for EEG-based emotion recognition. The Gaussian kernel enhances the EEG data separability in the transformed space, and the Laplacian prior controls the complexity of the learned LR regressor in the training process. The LORSAL algorithm was introduced to optimize the LR with Gaussian kernel and Laplacian prior for its low computational complexity. Various spectral power features in the frequency domain and features by combining the asymmetrical electrodes, PSD, DE, DASM, RASM, and DCAU, were extracted for the Delta, Theta, Alpha, Beta, Gamma and Total frequency bands using 256-point STFT from the segmented 1 s EEG epochs.

The experiments were conducted on the publicly available DEAP dataset, and the performance of the introduced LORSAL methods was compared with the NB, SVM, LR_L1, and LR_L2 classifiers. The experimental results showed that LORSAL presented the best accuracies of 77.17% and 77.03% in valence and arousal dimensions, respectively, on the DE features from the total frequency bands, while the SVM classifiers obtained the second-highest accuracies of 69.55% and 69.92%. The other evaluation metrics obtained by LORSAL, SVM, and NB were also tabulated in the paper. The introduced LORSAL method also presented the best Recall (76.79% and 77.03% in valence and arousal, respectively) and F1 (76.90% and 76.47% in valence and arousal, respectively). The previous experimental results showed the superiority of the introduced LORSAL method for EEG-based emotion recognition compared to the NB, SVM, LR_L1, and LR_L2 approaches.

This paper also showed an investigation of the critical frequency bands for EEG-based emotion recognition. In this study, the informative features are captured from different frequency bands: Delta, Theta, Alpha, Beta, Gamma, and Total. The previous neuroscience studies showed that specific frequency band ranges are associated with specific brain activities. For example, the EEG Alpha frequency bands are related to attentional processing, whereas the Beta bands are a reflection of emotional and cognitive processing. The experimental results showed that the LORSAL, SVM, and NB classifiers performed better on the Gamma and Beta frequency bands than other bands for different features. The comparison of the Fisher ratio also showed the effectiveness of Gamma and Beta bands in emotion recognition. The findings in this study are in accordance with the previous work about critical bands investigation [47,48].

Additionally, the effects of different features, PSD, DE, DASM, RASM, and DCAU, on the emotion classification results were also analyzed in this paper. Experimental results show that the compared approaches, LORSAL, SVM, and NB obtained superior precision metrics on the DE features over other features. This shows the effectiveness of the DE features in distinguishing low- and high-frequency energy in EEG sequences. Meanwhile, the DASM and DCAU features presented relatively ideal classification accuracies compared to the PSD features. It is noted that DASM and DCAU have the advantages of less time consumption for their lower dimensionality than PSD and DE.

For a more comprehensive analysis, -Table 5 showed a comparison of the introduced LORSAL methods, the other shallow classifiers, and the deep learning approaches for EEG-based emotion recognition of LV/HV and LA/HA on DEAP dataset. In single-trial classification by Koelstra et al. [27], the NB after feature selection using Fisher's linear discrimination, obtained the accuracies of 57.65% and 62.0% in valence and arousal dimensions. In [72], the Bayesian weighted-log-posterior function optimized with the perceptron convergence algorithm presented average precisions of 70.9% and 70.1% for valence and arousal. For within-subject emotion recognition of LV/HV and LA/HA, Atkinson et al. [73] presented the accuracies of 73.41% and 73.06% using minimum-Redundancy-Maximum-Relevance (mRMR) for feature selection. Rozgić et al. [74] performed classification using segment level decision fusion and presented precisions of 76.9% and 69.4% to discriminate LV/HV and LA/HA emotions. In the studies by Zheng et al. [48], the discriminative graph regularized extreme learning machine (GELM) with DE features achieved the highest average accuracies of 69.67% for 4-class classification in VA emotion space. The introduced LORSAL classifier presented ideal evaluation metrics for EEG emotion recognition, including the compared NB, SVM, LR_L1, and LR_L2 methods in the experiments.

Recently, deep learning (DL) methods have been used for EEG-based emotion classification [28,29]. In [75], a hybrid DL model combining CNN and RNN learned task-related features from grid-like EEG frames and achieved the accuracies of 72.06% and 74.12 for valence and arousal. The DNN and CNN models by Tripathi et al. [76] achieved the precisions of 75.78%, 73.12%, 81.41%, and 73.36% along valence and arousal dimensions, respectively. The classification accuracies for valence and arousal were over 85% using LSTM-RNN by Alhagry et al. [77], and over 87% using 3D-CNN by Salama et al. [78]. More recently, Chen et al. [33,34] have researched a lot on the combination of DL models and various features. As tabulated in Table 5, computer vision CNN (CVCNN), global spatial filter CNN (GSCNN), and global space local time filter (GSLTCNN) [33] presented obvious improvements with concatenating PSD, raw EEG features, and normalized EEG signals. In [34], the proposed hierarchical bidirectional gated recurrent unit (H-ATT-BGRU) network performed better on raw EEG signals than CNN and LSTM, and the obtained accuracies in valence and arousal dimensions were 67.9% and 66.5% for 2-class cross-subject emotion recognition. For more details about the DL architectures applied in the DEAP data, readers may refer to the literature [33,34,75–78]. Compared to traditional shallow methods, the DL schemes remove the signal pre-processing and feature extraction/selection progress, and are more suitable for affective representation [35,36]. However, the DL methods cannot reveal the relationship between emotional states and EEG signals for being like a black box [37].

However, more importantly, the training of DL networks is extremely time-consuming, which limits their practical applications in real-time emotion recognition [3]. Craik et al. [28] stated that from practical issues, the DL methods have problems of very long computation, and the vanishing/ exploding gradients, and their practical application need extra graphic processing unit (GPU). Roy et al. [29] pointed out that from a practical point-of-view, the hyperparameter search of a DL algorithm often takes up a lot of time for training. Additionally, Craik et al. [28] and Roy et al. [29] make comprehensive reviews on the recent DL schemes.

To illustrate the time efficiency, the average training time of the compared NB, SVM, MLR_L1, MLR_L2, and LORSAL methods are shown in Table 6. The average running time for STFT-based feature extraction is 68.15 s. In our experiment, all the programs are performed using on a computer with an Intel Core i5-4590 of 3.30 GHz and 8.00-GB RAM. LORSAL takes just no more than 4 s for training, and the computing time is in the same order as the compared traditional shallow methods. As mentioned earlier, the complexity of LORSAL is $O((L+1)^2K)$ for each quadratic problem, where L is the number of EEG epochs used for training and K is the number of emotion classes. As shown in Table 5, the time-consumptions of LORSAL on DE, PSD, DASM, RASM, and DCAU (with different dimensions 160, 160, 70, 70, and 55) are nearly the same. Given limited computational resources, or with portable devices, the introduced LORSAL algorithm has higher time efficiency than DL methods and can present better performance than the compared shallow methods.

Table 5. Comparison of the introduced LORSAL methods, the other shallow classifiers, and the deep learning approaches for EEG-based emotion recognition of LV/HV and LA/HA on DEAP dataset.

Classifier			Valence	Arousal	Description
NB		by Koelstra et al. [27]	57.6	62.0	2-class classification for valence and arousal, and within-subject emotion recognition.
Bayesian weighted-log-posterior		by Yoon et al. [72]	70.9	70.1	
SVM+mRMR		by Atkinson et al. [73]	73.41	73.06	
Segment level decision fusion		by Rozgić et al. [74]	76.9	68.4	
CNN+RNN		by Li et al. [75]	72.06	74.12	
DNN		by Tripathi et al. [76]	75.78	73.12	
CNN			81.41	73.36	
LSTM-RNN		by Alhagry et al. [77]	85.65	85.45	
3D-CNN		by Salama et al. [78]	87.44	88.49	
GELM		by Zheng et al. [48]	69.7		4-class classification in VA space.
SVM	+Raw	by Chen et al. [33]	0.5590	0.7525	2-class classification for valence and arousal, and within-subject emotion recognition, and AUC (Area Under ROC Curve) used for evaluation.
	+Norm		0.5591	0.5590	
	+PSD		0.7596	0.5531	
	+PSD+Raw		0.9234	0.9462	
	+PSD+Norm		0.7460	0.7353	
CVCNN	+Raw		0.6221	0.6012	
	+Norm		0.6551	0.6176	
	+PSD		0.9307	0.88.51	
	+PSD+Raw		0.9933	0.9988	
	+PSD+Norm		1.00	1.00	
GSCNN	+Raw		0.6242	0.5902	
	+Norm		0.6394	0.5987	
	+PSD		0.8875	0.8802	
	+PSD+Raw		0.9933	0.9930	
	+PSD+Norm		1.00	1.00	
GSLTCNN	+Raw		0.6717	0.6175	
	+Norm		0.6350	0.5670	
	+PSD		0.8523	0.8390	
	+PSD+Raw		0.9946	0.9958	
	+PSD+Norm		1.00	1.00	
CNN	+Raw	by Chen et al. [34]	57.2	56.3	2-class classification for valence and arousal, and cross-subject emotion recognition
LSTM			63.7	61.9	
H-ATT-BGRU			67.9	66.5	
NB	+DE	in our study	61.04	58.46	2-class classification for valence and arousal, and within-subject emotion recognition
SVM			69.55	69.92	
MLR_L1			33.69	43.10	
MLR_L2			33.81	43.25	
LORSAL			77.17	77.03	

Table 6. Running time of the compared NB, SVM, MLR_L1, MLR_L2, and LORSAL methods in terms of second and the average time-consumption of feature extraction is 68.15 s.

Classifier	Valence					Arousal				
	PSD	DE	DASM	RASM	DCAU	PSD	DE	DASM	RASM	DCAU
NB	4.37	4.36	1.91	1.92	1.51	4.42	4.41	1.93	1.93	3.44
SVM	4.53	4.37	2.34	2.47	2.04	4.26	4.16	2.26	2.34	3.45
MLR_L1	0.14	0.13	0.04	0.04	0.03	0.17	0.15	0.04	0.04	0.08
MLR_L2	0.12	0.11	0.04	0.04	0.03	0.13	0.11	0.04	0.04	0.07
LORSAL	3.75	3.89	3.87	3.69	3.85	3.74	3.86	3.86	3.69	3.87

5. Conclusions and Future Work

This paper systematically investigates the introduced LORSAL algorithm for the EEG-based emotion class. Additionally, the critical frequency bands, Delta, Theta, Alpha, Beta, Gamma, and the effectiveness of different features, PSD, DE, DASM, RASM, and DCAU, on emotion recognition are also analyzed. The LORSAL classifier performs better than the compared shallow methods and has the superiority of time efficiency compared to the recent DL approaches.

The performance and application of LORSAL-based emotion recognition should be further researched in future work. More informative and representative features can be used in LORSAL. As shown in Table 5, in the research by Chen et al. [33], SVM achieved higher values of AUC (Area Under ROC Curve), 0.9234 and 0.9426, for classifying LV/HV and LA/HA emotions by concatenating PSD and raw pre-processed EEG signals than with other features. We will try to integrate different features to train the LORSAL classifier. The future attempts include the application of LORSAL for 4-class emotion classification in VA space, as the studies in [48]. A further comparison of LORSAL and DL methods, and the combination of their advantages in feature extraction and avoiding overfitting will be investigated. Future work could also include applying LORSAL on multimodal information, e.g., fNIRS, and other physiological signals in brain activity analysis [79,80].

Author Contributions: Conceptualization, C.P. and C.S.; methodology, C.P. and C.S.; software, C.P. and C.S; validation, C.P.; formal analysis, C.P. and C.S; investigation, C.P.; writing—original draft preparation, C.P.; writing—review and editing, C.S.; supervision, H.M., J.L. and X.G.; funding acquisition, C.P., C.S., and H.M. All authors have read and agreed to the published version of the manuscript.

Funding: This research was supported by the National Natural Science Foundation of China under Grant 61902313, the Fundamental Research Funds for the Central Universities, Xidian University, No. RW190110, and the Construction Project Achievement of College Counselor Studio of Shaanxi Province: Reach Perfection with Morality Studio.

Acknowledgments: The authors would like to thank J. Li for providing the source codes of the LORSAL algorithm on the websites (http://www.lx.it.pt/~{}jun/).

Conflicts of Interest: The authors declare no conflict of interest.

References

1. Picard, R.W. *Affective Computing*; MIT Press: Cambridge, MA, USA, 1997.
2. Levenson, R.W. The autonomic nervous system and emotion. *Emotion Rev.* **2014**, *6*, 100–112. [CrossRef]
3. Bota, P.J.; Wang, C.; Fred, A.L.; Da Silva, H.P. A review, current challenges, and future possibilities on emotion recognition using machine learning and physiological signals. *IEEE Access* **2019**, *7*, 140990–141020. [CrossRef]
4. Shu, L.; Xie, J.; Yang, M.; Li, Z.; Li, Z.; Liao, D.; Xu, X.; Yang, X. A review of emotion recognition using physiological signals. *Sensors* **2018**, *8*, 2074. [CrossRef]
5. Cannon, W.B. The James-Lange theory of emotions: A critical examination and an alternative theory. *Am. J. Psychol.* **1927**, *39*, 106–124. [CrossRef]
6. Ayaz, H.; Curtin, A.; Mark, J.; Kraft, A.; Ziegler, M. Predicting Future Performance based on Current Brain Activity: An fNIRS and EEG Study. In Proceedings of the 2019 IEEE International Conference on Systems, Man and Cybernetics (SMC), Bari, Italy, 6–9 October 2019; pp. 3925–3930.

7. Saadati, M.; Nelson, J.; Ayaz, H. Convolutional Neural Network for Hybrid fNIRS-EEG Mental Workload Classification. In Proceedings of the International Conference on Applied Human Factors and Ergonomics, Washington, DC, USA, 24–28 July 2019; pp. 221–232.
8. Jiao, Z.; Gao, X.; Wang, Y.; Li, J.; Xu, H. Deep Convolutional Neural Networks for mental load classification based on EEG data. *Pattern Recognit.* **2018**, *76*, 582–595. [CrossRef]
9. Sargent, A.; Heiman-Patterson, T.; Feldman, S.; Shewokis, P.A.; Ayaz, H. *Mental Fatigue Assessment in Prolonged BCI Use Through EEG and fNIRS.-Neuroergonomics*; Academic Press: Cambridge, MA, USA, 2018.
10. Abdul, A.; Chen, J.; Liao, H.Y.; Chang, S.H. An emotion-aware personalized music recommendation system using a convolutional neural networks approach. *Appl. Sci.* **2018**, *8*, 1103. [CrossRef]
11. Jiao, Z.; You, H.; Yang, F.; Li, X.; Zhang, H.; Shen, D. Decoding EEG by visual-guided deep neural networks. In Proceedings of the 28th International Joint Conference on Artificial Intelligence, Macao, China, 10–16 August 2019; pp. 1387–1393.
12. Ren, Z.; Li, J.; Xue, X.; Li, X.; Yang, F.; Jiao, Z.; Gao, X. Reconstructing Perceived Images from Brain Activity by Visually-guided Cognitive Representation and Adversarial Learning. *arXiv* **2019**, arXiv:1906.12181.
13. Lotte, F.; Congedo, M.; Lécuyer, A.; Lamarche, F.; Arnaldi, B. A review of classification algorithms for EEG-based brain–computer interfaces. *J. Neural Eng.* **2007**, *4*, R1–R13. [CrossRef] [PubMed]
14. Lotte, F.; Bougrain, L.; Cichocki, A.; Clerc, M.; Congedo, M.; Rakotomamonjy, A.; Yger, F. A review of classification algorithms for EEG-based brain–computer interfaces: A 10 year update. *J. Neural Eng.* **2018**, *15*, 031005. [CrossRef] [PubMed]
15. Russell, J.A. A circumplex model of affect. *J. Personal. Soc. Psychol.* **1980**, *39*, 1161–1178. [CrossRef]
16. Kim, M.-K.; Kim, M.; Oh, E.; Kim, S.-P. A review on the computational methods for emotional state estimation from the human EEG. *Comput. Math. Methods Med.* **2013**, *2013*, 1–13. [CrossRef] [PubMed]
17. Cacioppo, J.T. Feelings and emotions: Roles for electrophysiological markers. *Biol. Psychol.* **2004**, *67*, 235–243. [CrossRef] [PubMed]
18. Sanei, S.; Chambers, J. *EEG Signal Processing*; Wiley: New York, NY, USA, 2007.
19. Al-Nafjan, A.; Hosny, M.; Al-Ohali, Y.; Al-Wabil, A. Review and classification of emotion recognition based on EEG brain-computer interface system research: A systematic review. *Appl. Sci.* **2017**, *7*, 1239. [CrossRef]
20. Kroupi, E.; Vesin, J.M.; Ebrahimi, T. Subject-independent odor pleasantness classification using brain and peripheral signals. *IEEE Trans. Affect. Comput.* **2016**, *7*, 422–434. [CrossRef]
21. Zhang, J.H.; Chen, M.; Zhao, S.K.; Hu, S.Q.; Shi, Z.G.; Cao, Y. Relieff-based EEG sensor selection methods for emotion recognition. *Sensors* **2016**, *16*, 1558. [CrossRef]
22. Chew, L.H.; Teo, J.; Mountstephens, J. Aesthetic preference recognition of 3d shapes using EEG. *Cogn. Neurodyn.* **2016**, *10*, 165–173. [CrossRef]
23. Tang, J.; Alelyani, S.; Liu, H. Feature Selection for Classification: A Review. In *Data Classification: Algorithms and Applications*; CRC Press: Boca Raton, FL, USA, 2014; pp. 37–64.
24. Chao, G.; Luo, Y.; Ding, W. Recent advances in supervised dimension reduction: A Survey. *Mach. Learn. Knowl. Extr.* **2019**, *1*, 20. [CrossRef]
25. Lin, Y.-P.; Wang, C.-H.; Wu, T.-L.; Jeng, S.-K.; Chen, J.-H. EEG-based emotion recognition in music listening: A comparison of schemes for multiclass support vector machine. In Proceedings of the IEEE International Conference on Acoustics, Speech and Signal Processing, Taipei, Taiwan, 19–24 April 2009; pp. 489–492.
26. Horlings, R.; Datcu, D.; Rothkrantz, L.J.M. Emotion recognition using brain activity. In Proceedings of the 9th International Conference on Computer Systems and Technologies and Workshop for PhD students in Computing, Gabrovo, Bulgaria, 12–13 June 2008; ACM: New York, NY, USA, 2008.
27. Koelstra, S.; Muhl, C.; Soleymani, M.; Lee, J.S.; Yazdani, A.; Ebrahimi, T.; Pun, T.; Nijholt, A.; Patras, I. DEAP: A database for emotion analysis; using physiological signals. *IEEE Trans. Affect. Comput.* **2012**, *3*, 18–31. [CrossRef]
28. Craik, A.; He, Y.; Contreras-Vidal, J.L. Deep learning for electroencephalogram (EEG) classification tasks: A review. *J. Neural Eng.* **2019**, *16*, 031001. [CrossRef]
29. Roy, Y.; Banville, H.; Albuquerque, I.; Gramfort, A.; Falk, T.H.; Faubert, J. Deep learning-based electroencephalography analysis: A systematic review. *J. Neural Eng.* **2019**, *16*, 051001. [CrossRef]
30. Zheng, W.L.; Zhu, J.Y.; Peng, Y.; Lu, B.L. EEG-based emotion classification using deep belief networks. In Proceedings of the 2014 IEEE International Conference on Multimedia and Expo (ICME), Chengdu, China, 14–18 July 2014; pp. 1–6.

31. Zheng, W.L.; Guo, H.T.; Lu, B.L. Revealing critical channels and frequency bands for emotion recognition from EEG with deep belief network. In Proceedings of the 2015 7th International IEEE/EMBS Conference on Neural Engineering (NER), Montpellier, France, 22–24 April 2015; pp. 154–157.
32. Yang, Y.; Wu, Q.; Qiu, M.; Wang, Y.; Chen, X. Emotion recognition from multi-channel EEG through parallel convolutional recurrent neural network. In Proceedings of the International Joint Conference on Neural Networks, Rio, Brasil, 8–13 July 2018; pp. 1–7.
33. Chen, J.X.; Zhang, P.W.; Mao, Z.J.; Huang, Y.F.; Jiang, D.M.; Zhang, Y.N. Accurate EEG-based emotion recognition on combined features using deep convolutional neural networks. *IEEE Access* **2019**, *7*, 44317–44328. [CrossRef]
34. Chen, J.X.; Jiang, D.M.; Zhang, Y.N. A hierarchical bidirectional GRU model with attention for EEG-based emotion classification. *IEEE Access* **2019**, *7*, 118530–118540. [CrossRef]
35. Martinez, H.P.; Bengio, Y.; Yannakakis, G.N. Learning deep physiological models of affect. *IEEE Comput. Intell. Mag.* **2013**, *8*, 20–33. [CrossRef]
36. Chen, J.X.; Mao, Z.J.; Yao, W.X.; Huang, Y.F. EEG-based biometric identification with convolutional neural network. *Multimed. Tools Appl.* **2019**, 1–21. [CrossRef]
37. Lee, J.; Yoo, S.K. Design of user-customized negative emotion classifier based on feature selection using physiological signal sensors. *Sensors* **2018**, *18*, 4253. [CrossRef]
38. Hosmer, D.W.; Lemeshow, S.; Sturdivant, R.X. *Applied Logistic Regression*; John Wiley & Sons: Hoboken, NJ, USA, 2013.
39. Alkan, A.; Koklukaya, E.; Subasi, A. Automatic seizure detection in EEG using logistic regression and artificial neural network. *J. Neurosci. Methods* **2005**, *148*, 167–176. [CrossRef]
40. Subasi, A.; Ercelebi, E. Classification of EEG signals using neural network and logistic regression. *Comput. Methods Programs Biomed.* **2005**, *78*, 87–99. [CrossRef]
41. Tomioka, R.; Aihara, K.; Müller, K.R. Logistic regression for single trial EEG classification. In *Advances in Neural Information Processing Systems 19, Proceedings of the Twentieth Annual Conference on Neural Information Processing Systems, Vancouver, BC, Canada, 4–7 December 2006*; MIT Press: Cambridge, MA, USA, 2006.
42. Ezzyat, Y.; Kragel, J.E.; Burke, J.F.; Levy, D.F.; Lyalenko, A.; Wanda, P.; O'Sullivan, L.; Hurley, K.B.; Busygin, S.; Pedisich, I.; et al. Direct brain stimulation modulates encoding states and memory performance in humans. *Curr. Biol.* **2017**, *27*, 1–8. [CrossRef]
43. Arora, A.; Lin, J.J.; Gasperian, A.; Maldjian, J.; Stein, J.; Kahana, M.; Lega, B. Comparison of logistic regression, support vector machines, and deep learning classifiers for predicting memory encoding success using human intracranial EEG recordings. *J. Neural Eng.* **2018**, *15*, 066028. [CrossRef]
44. Krishnapuram, B.; Carin, L.; Figueiredo, M.A.T.; Member, S. Sparse multinomial logistic regression: Fast algorithms and generalization bounds. *IEEE Trans. Pat. Anal. Mach. Intell.* **2005**, *27*, 957–968. [CrossRef]
45. Bioucas-Dias, J.; Figueiredo, M. *Logistic Regression via Variable Splitting and Augmented Lagrangian Tools*; Instituto Superior Técnico: Lisboa, Portugal, 2009.
46. Li, J.; Bioucas-Dias, J.M.; Plaza, A. Hyperspectral image segmentation using a new Bayesian approach with active learning. *IEEE Trans. Geosci. Remote Sens.* **2011**, *49*, 3947–3960. [CrossRef]
47. Zheng, W.L.; Lu, B.L. Investigating critical frequency bands and channels for EEG-based emotion recognition with deep neural networks. *IEEE Trans. Auton. Ment. Dev.* **2015**, *7*, 162–175. [CrossRef]
48. Zheng, W.L.; Zhu, J.Y.; Lu, B.L. Identifying stable patterns over time for emotion recognition from EEG. *IEEE Trans. Affect. Comput.* **2019**, *10*, 417–429. [CrossRef]
49. Bradley, M.M.; Lang, P.J. Measuring emotion: The self-assessment manikin and the semantic differential. *J. Behav. Ther. Exp. Psychiatry* **1994**, *25*, 49–59. [CrossRef]
50. Jenke, R.; Peer, A.; Buss, M. Feature extraction and selection for emotion recognition from EEG. *IEEE Trans. Affect. Comput.* **2014**, *5*, 327–339. [CrossRef]
51. Balconi, M.; Mazza, G. Brain oscillations and BIS/BAS (behavioral inhibition/activation system) effects on processing masked emotional cues: ERS/ERD and coherence measures of alpha band. *Int. J. psychophysiol.* **2009**, *74*, 58–65. [CrossRef] [PubMed]
52. Bos, D.O. EEG-based emotion recognition The Influence of Visual and Auditory Stimuli. *Emotion* **2006**, *1359*, 667–670.

53. Chanel, G.; Karim, A.-A.; Thierry, P. Valence-arousal evaluation using physiological signals in an emotion recall paradigm. In Proceedings of the IEEE International Conference on Systems, Man and Cybernetics, Montreal, QC, Canada, 7–10 October 2007.
54. Lin, Y.P.; Wang, C.H.; Jung, T.P.; Wu, T.L.; Jeng, S.K.; Duann, J.R.; Chen, J.H. EEG-based emotion recognition in music listening. *IEEE Trans. Biomed. Eng.* **2010**, *57*, 1798–1806.
55. Shi, L.; Jiao, Y.; Lu, B. Differential entropy feature for EEG-based vigilance estimation. In Proceedings of the 35th Annual International Conference of the IEEE Engineering in Medicine and Biology Society (EMBC), Osaka, Japan, 3–7 July 2013; pp. 6627–6630.
56. Duan, R.; Zhu, J.; Lu, B. Differential entropy feature for EEG-based emotion classification. In Proceedings of the 6th International IEEE/EMBS Conference on Neural Engineering (NER), San Diego, CA, USA, 6–8 November 2013; pp. 81–84.
57. Gibbs, J.W. *Elementary Principles in Statistical Mechanics—Developed with Especial Reference to the Rational Foundation of Thermodynamics*; C. Scribner's Sons: New York, NY, USA, 1902.
58. Davidson, R.; Fox, N. Asymmetrical brain activity discriminates between positive and negative stimuli infants. *Science* **1982**, *218*, 1235–1237. [CrossRef]
59. Lin, Y.P.; Yang, Y.H.; Jung, T.P. Fusion of electroencephalographic dynamics and musical contents for estimating emotional responses in music listening. *Front. Neurosci.* **2014**, *8*, 94. [CrossRef]
60. Li, J.; Bioucas-Dias, J.M.; Plaza, A. Semi-supervised hyperspectral image segmentation using multinomial logistic regression with active learning. *IEEE Trans. Geosci. Remote Sens.* **2010**, *48*, 4085–4098.
61. Hunter, D.R.; Lange, K. A tutorial on MM algorithms. *Amer. Statistician* **2004**, *58*, 30–37. [CrossRef]
62. Borges, J.S.; Bioucas-Dias, J.M.; Marçal, A.R.S. Fast Sparse Multinomial Regression Applied to Hyperspectral Data. In Proceedings of the Third International Conference on Image Analysis and Recognition—Volume Part II, Póvoa de Varzim, Portugal, 18–20 September 2006; Springer: Berlin, Germany, 2006.
63. Fan, R.E.; Chang, K.W.; Hsieh, C.J.; Wang, X.R.; Lin, C.J. LIBLINEAR: A library for large linear classification. *J. Mach Learn. Res.* **2008**, *9*, 1871–1874.
64. Chang, C.-C.; Lin, C.-J. LIBSVM: A library for support vector machines. *ACM Trans. Intell. Syst. Technol.* **2011**, *2*, 27. [CrossRef]
65. Mao, K.Z. RBF neural network center selection based on Fisher ratio class separability measure. *IEEE Trans. Neural Netw.* **2002**, *13*, 1211–1217. [CrossRef] [PubMed]
66. Wang, L. Feature selection with kernel class separability. *IEEE Trans. Pat. Anal. Mach. Intell.* **2008**, *30*, 1534–1546. [CrossRef] [PubMed]
67. Pan, C.; Gao, X.; Wang, Y.; Li, J. Markov random fields integrating adaptive interclass-pair penalty and spectral similarity for hyperspectral image classification. *IEEE Trans. Geosci. Remote Sens.* **2018**, *57*, 2520–2534. [CrossRef]
68. Ray, W.J.; Cole, H.W. EEG alpha activity reflects attentional demands, and beta activity reflects emotional and cognitive processes. *Science* **1985**, *228*, 750–752. [CrossRef]
69. Klimesch, W.; Doppelmayr, M.; Russegger, H.; Pachinger, T.; Schwaiger, J. Induced alpha band power changes in the human EEG and attention. *Neurosci. Lett.* **1998**, *244*, 73–76. [CrossRef]
70. Onton, J.; Makeig, S. High-frequency broadband modulations of electroencephalographic spectra. *Front. Neurosci.* **2009**, *3*, 61. [CrossRef] [PubMed]
71. Mu, L.; Lu, B.-L. Emotion classification based on gamma-band EEG. In Proceedings of the Annual International Conference of the IEEE, Minneapolis, MN, USA, 3–6 September 2009; pp. 1323–1326.
72. Yoon, H.J.; Chung, S.Y. EEG-based emotion estimation using Bayesian weighted-log-posterior function and perceptron convergence algorithm. *Comput. Biol. Med.* **2013**, *43*, 2230–2237. [CrossRef] [PubMed]
73. Atkinson, J.; Campos, D. Improving BCI-based emotion recognition by combining EEG feature selection and kernel classifiers. *Expert Syst. Appl.* **2016**, *47*, 35–41. [CrossRef]
74. Rozgic, V.; Vitaladevuni, S.N.; Prasad, R. Robust the EEG emotion classification using segment level decision fusion. In Proceedings of the IEEE Conference of Acoustics, Speech, and Signal Processing (ICASSP), Vancouver, BC, Canada, 26–31 May 2013; pp. 1286–1290.
75. Li, X.; Song, D.; Zhang, P.; Yu, G.; Hu, B. Emotion recognition from multi-channel the EEG data through convolutional recurrent neural network. In Proceedings of the IEEE International Conference on Bioinformatics and Biomedicine (BIBM), Shenzhen, China, 15–18 December 2016; pp. 352–359.

76. Tripathi, S.; Acharya, S.; Sharma, R.D.; Mittal, S.; Bhattacharya, S. Using deep and convolutional neural networks for accurate emotion classification on DEAP dataset. In Proceedings of the Twenty-Ninth AAAI Conference on Innovative Applications (IAAI-17), Austin, TX, USA, 25–30 January 2015; pp. 4746–4752.
77. Alhagry, S.; Fahmy, A.A.; El-Khoribi, R.A. Emotion recognition based on EEG using LSTM recurrent neural network. *Emotion* **2017**, *8*, 355–358. [CrossRef]
78. Salama, E.S.; El-Khoribi, R.A.; Shoman, M.E.; Shalaby, M.A.W. EEG-based emotion recognition using 3D convolutional neural networks. *Int. J. Adv. Comput. Sci. Appl.* **2018**, *9*, 329–337. [CrossRef]
79. Liu, Y.; Ayaz, H.; Shewokis, P.A. Multisubject "learning" for mental workload classification using concurrent EEG, fNIRS, and physiological measures. *Front. Hum. Neurosci.* **2017**, *11*, 389. [CrossRef] [PubMed]
80. Saadati, M.; Nelson, J.; Ayaz, H. Multimodal fNIRS-EEG Classification Using Deep Learning Algorithms for Brain-Computer Interfaces Purposes. In Proceedings of the International Conference on Applied Human Factors and Ergonomics, Washington, DC, USA, 24–28 July 2019; pp. 209–220.

applied sciences

MDPI

Article

Deep Learning for EEG-Based Preference Classification in Neuromarketing

Mashael Aldayel [1,2,*,†], Mourad Ykhlef [1,†] and Abeer Al-Nafjan [3,†]

1 Information System Department, College of Computer and Information Sciences, King Saud University, Riyadh 11543, Saudi Arabia; ykhlef@ksu.edu.sa
2 Information Technology Department, College of Computer and Information Sciences, King Saud University, Riyadh 11543, Saudi Arabia
3 Computer Science Department, College of Computer and Information Sciences, Imam Muhammad bin Saud University, Riyadh 11432, Saudi Arabia; annafjan@imamu.edu.sa
* Correspondence: maldayel@ksu.edu.sa
† Current address: Riyadh 11432, Saudi Arabia.

Received: 14 January 2020; Accepted: 19 February 2020; Published: 24 February 2020

Featured Application: This article presents an application of deep learning in preference detection performed using EEG-based BCI.

Abstract: The traditional marketing methodologies (e.g., television commercials and newspaper advertisements) may be unsuccessful at selling products because they do not robustly stimulate the consumers to purchase a particular product. Such conventional marketing methods attempt to determine the attitude of the consumers toward a product, which may not represent the real behavior at the point of purchase. It is likely that the marketers misunderstand the consumer behavior because the predicted attitude does not always reflect the real purchasing behaviors of the consumers. This research study was aimed at bridging the gap between traditional market research, which relies on explicit consumer responses, and neuromarketing research, which reflects the implicit consumer responses. The EEG-based preference recognition in neuromarketing was extensively reviewed. Another gap in neuromarketing research is the lack of extensive data-mining approaches for the prediction and classification of the consumer preferences. Therefore, in this work, a deep-learning approach is adopted to detect the consumer preferences by using EEG signals from the DEAP dataset by considering the power spectral density and valence features. The results demonstrated that, although the proposed deep-learning exhibits a higher accuracy, recall, and precision compared with the k-nearest neighbor and support vector machine algorithms, random forest reaches similar results to deep learning on the same dataset.

Keywords: neuromarketing; brain computer interface (BCI); consumer preferences; EEG signal; deep learning; deep neural network (DNN)

1. Introduction

Neuromarketing is an emerging field that links the cognitive and affective sides of the consumer behavior by using neuroscience. Since its origin in 2002, this field has rapidly achieved credibility among the advertising and marketing specialists, and many such specialists are adopting neuromarketing strategies [1].

Neuromarketing can assist marketers in understanding how a consumer's brain evaluates the diverse brands and recognizing the factors that affect the consumers' choices when purchasing products. Neuromarketing research has demonstrated that people do not always recognize what happens in their unconscious brains. Furthermore, it has been demonstrated that people are not always explicit in their preferences or intentions [2].

The use of traditional marketing tools, such as interviews and questionnaires, to assess consumer preferences, needs, and buying intentions can lead to the generation of biased or incorrect conclusions [3,4]. Similarly, an oral expression of preferences can produce conscious or unconscious biases. It is difficult to extract the consumer preferences directly through choices, owing to the high product costs, ethical caution considerations, or the product not having been invented at the time of evaluation [3]. These elements highlight a contradiction in the users' opinions during the usability assessments and their actual opinions, feelings, and senses regarding the use of a product [4].

Therefore, neuromarketing requires more effective methodological alternatives to evaluate the consumer behavior. Novel neuroimaging procedures provide an effective approach to study consumer behavior. Such methods ultimately help marketers examine the consumers' brains to obtain valuable insights into the subconscious procedures underlying successful or failed marketing messages. This information is obtained by eliminating the primary problem in traditional advertising research, that is, trusting people; in particular, people should be trusted, whether they are consumers or workers who report on how the consumers are influenced by a specific part of an advertisement [1].

Brain–computer interfaces (BCIs) are promising neuroimaging tools in neuromarketing. This technology allows the users to communicate effectively with computer systems. A BCI does not require the use of any external devices or muscle interference to produce commands [5]. Furthermore, a BCI employs voluntarily generated user brain activity to control a system through signals, which provides the ability to communicate or interact with the nearby environment. Electroencephalography (EEG) is one of the main instruments used to examine brain activity. The EEG technique is the only practical, versatile, affordable, portable, non-invasive BCI to perform repetitive, real-time analysis of brain interactions in high temporal resolution [5–7].

Therefore, in the present research study, EEG was adopted as the input brain signal for a BCI system. Using classification algorithms, BCIs can be used as neural measures to distinguish the preference patterns from brain signals and translate them into actions to promote a product. In addition, the performance of a deep neural network (DNN) was implemented and examined to model a benchmark dataset for the preference classification.

The main objective of this research was to deeply investigate EEG-based preference recognition in neuromarketing to enhance the accuracy of classification prediction by comparing the performance of deep-learning with other conventional classification algorithms, such as support vector machine (SVM), random forest (RF), and k-nearest neighbors (KNN).

2. Background

This section provides a review of the main concepts used in this research: neuromarketing, BCI, and EEG.

2.1. Neuromarketing

Traditional marketing approaches include surveys, interviews, questionnaires, and focus groups, in which consumers openly and consciously report their experiences and opinions. However, such traditional approaches cannot evaluate the unconscious side of the consumer behavior. Neuroscience has the potential to discern the unconscious motivations that influence the act of making choices. It has been reported that approximately 90% of data are processed subconsciously in the human mind [8]. In the field of neuromarketing and consumer neuroscience, the evaluation of the subconscious activities exposes the true preferences of consumers more accurately than traditional marketing research does. Furthermore, neuromarketing can reveal information regarding the consumer preferences/ratings that cannot be accurately determined through the traditional methods. This is because subconscious opinions play a key role in consumer decision-making. Traditional market research approaches fail to assess the subconscious activities in the consumer brain, which leads to an inequality between the results of the traditional market research and the real behavior of the consumers at the points of purchase [8].

The term "neuromarketing" is derived by combining the prefix "neuro" and the term 'marketing", indicating the integration of two study areas: neuroscience and marketing [1]. Neuroscience is a field that examines the facets of the brain at the biological level and from a psychological perspective [2]. In addition, neuroscience has significantly enlightened the field of marketing, and the interaction between these fields assists in intuiting the consumer behavior [8].

The term "neuromarketing" began to emerge organically around 2002. At that time, a few corporations, such as Brighthouse and SalesBrain, began offering neuromarketing studies and consultations, motivating the application of technology and knowledge from cognitive neuroscience to the field of marketing. Neuromarketing values the study of the consumer behavior from a psychological perspective [1]. Recently, several high-profile companies have begun exploiting neuromarketing approaches to assess the advertisements before introducing products to consumers [9]. This neuromarketing approach has gradually gained favor with brand executives in major corporations, such as Coca-Cola and Campbell's [10].

Neuromarketing researchers aim to use neuroscientific procedures to exploit consumer behavior (i.e., requests, needs, and preferences) when shoppers purchase goods. This factor represents the researchers' primary motivation for examining the consumers' sensorimotor, mental, and effective feedback for products and advertisements through various modalities [9]. There are several neuromarketing modalities besides BCI, such as eye-tracking, galvanic skin response, skin conductance, facial coding, and facial electromyography. Each modality records different neural measure [11]. Eye tracking is used to determine eye locations and eye movement to grasp the consumer's attention and natural responses to marketing stimuli. Galvanic skin response measures the moisture activity, which is related to the emotional state. Electromyography is used to evaluate the physiological features of facial muscles. Facial coding measures emotional states through facial expressions.

2.2. BCIs

BCIs are some of the most promising neuroimaging technologies in the neuromarketing domain. This technology helps facilitate effective communication between the users and computer systems. BCIs do not require any nerves, muscles, or movement interferences to issue a command [5] and employ the voluntarily-generated user brain activity to control a system through signals to communicate or interact with the nearby environment. Such environments can include wheelchairs, artificial arms/hands, and entertainment applications that involve skillful visualization, digital painting, and game playing [6].

BCI systems have contributed to numerous fields, including manufacturing, education, marketing, smart transportation, biomedical engineering, clinical neurology, and neuroscience [5,12]. A BCI system includes an input (i.e., the user's mental activity), output (i.e., states or commands), a decoder component between the input and output, and a protocol that regulates the beginning, offset, and timing of the action [6]. The BCI research is expected to lead to an approach in which the brain signals are operated to aid people in interaction actions [6].

In a BCI, the brain signals require processing in non-clinical situations, which corresponds to a new challenge in computational neuroscience research. Currently, most of the application-oriented BCI research is focused on endowing users—not only disabled people—the ability to control systems or sensors with various environments [6].

Different neuroimaging techniques can be recorded with non-invasive BCI, such as EEG, fNIRS, fMRI, PET, MEG, SST, and TMS. EEG has better temporal resolution than fNIRS, which is a relatively new neuroimaging technique in neuromarketing research. However, recent fNIRS research is still in the substantial validation phase [13,14]. EEG is most commonly used in neuromarketing research due to its advantages that are detailed in the next subsection.

2.3. EEG

The EEG is a widely used tool that examines the brain activity. The electrical activity is recorded on the scalp by evaluating the voltage variations from neurons firing in the brain. These electrical activities are logged over a period of time using several electrodes positioned on the scalp directly above the cortex. The electrodes are connected in a hat-like device [5,7]. The EEG has the following key benefits: it is non-invasive, portable, cost effective, and relatively simple to use, and it has an exceptional temporal resolution (up to milliseconds). However, the signal-to-noise ratio and spatial resolution are restricted compared with those of other techniques. Nevertheless, EEG is considered to be the only practical, non-invasive BCI input to realize a repetitive, real-time brain interactive analysis [5–7]. Therefore, EEG was selected as the input brain signal for the BCI in this research.

The international 10-20 system is a method used to name electrodes based on their location on the scalp. The approach relates information pertaining to the inter-electrode space, specifically, 10–20% of the front-to-back or right-to-left of the scalp boundaries. In other words, the distance between the nearby electrodes is either 10% or 20% of the scalp diameter, as depicted in Figure 1. The 10-20 standard has been frequently used across diverse EEG systems to increase the dependability of the signals and decrease the signal-to-noise ratio [5,7].

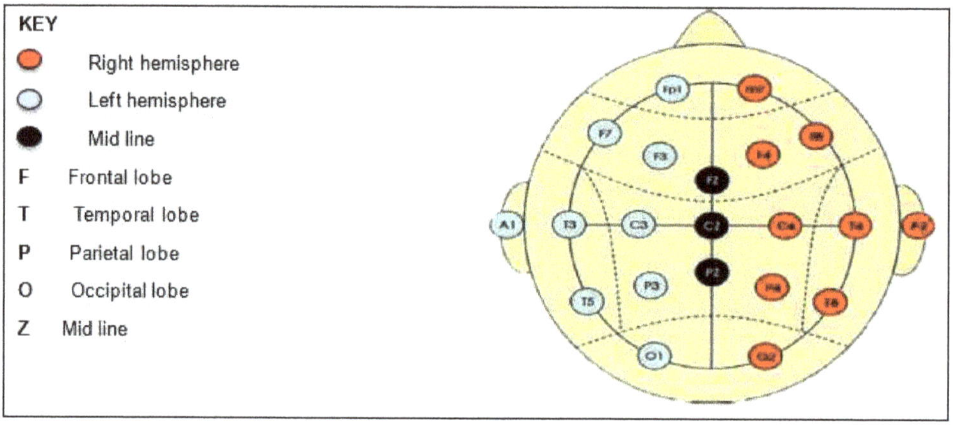

Figure 1. The international 10-20 system.

EEG Signals

The brain produces abundant neural activity, which can be captured as EEG signals for the BCI. These neural activities consist of two types: (1) rhythms; and (2) transient activities. The EEG activity can be further categorized on the basis of these types of activities [6,7].

1. **Rhythms**:

   Rhythms, neural oscillations, or brainwaves are repetitive forms of neural activity. The rhythms are measures of collective synaptic, neuronal, and axonal activities of the neuronal sets. The EEG activity is characterized by separating the frequencies into bands, denoted as delta, theta, alpha, beta, gamma, and mu rhythms. Table 1 presents the details of the EEG rhythms, ranges of frequency, amplitude, and shape, as well as the brain regions in which these activities are the most common along with the events usually associated with the type of band [6,7].

   These frequency bands have been linked to affective reactions. The theta band in the front-center of the brain reflects the emotional processing when a consumer looks at a product. The alpha band on the prefrontal cortex differentiates between the positive and negative emotional valences. The beta band is correlated with the alterations during affective arousal. Finally, the gamma band is largely associated with the arousal effects [15].

Table 1. Categorization of the EEG rhythms based on their frequency.

Rhythms	Frequency (Hz)	Amplitude (μV)	Brain Region	State of Mind
δ (Delta)	< 4	50 – 100	Central region	Deep sleep
θ (Theta)	4 – 8	<100	Central region	Drowsiness, first step of sleep
μ (Mu)	Approximately 10	< 50	Somatosensory cortex	Movement
α (Alpha)	8 – 13	< 10	Posterior brain	Opening the and focusing attention
β (Beta)	13 – 22	< 20	Central region	High state of wakefulness, alert and focused
γ (Gamma)	22 – 30	< 2	Highly localized	Higher mental activity, including perception and consciousness

2. **Transient activities**

The transient activities or field potentials replicate the action potentials of certain neurons in a manner similar to spikes. These spikes can be recognized by their position, frequency, amplitude, shape, recurrence, and operational properties. The event-related potentials (ERPs) and event-related spectral perturbations (ERSPs) are common types of transient activities [3,16].

ERP is the most common spike and arises as a reaction to a specific event or stimulus. These spikes have extremely small amplitudes. Consequently, the EEG samples must be averaged over many iterations to uncover the ERPs and eliminate noise fluctuations [16]. Table 2 presents the common ERPs used in the neuromarketing research. ERSPs compute the reaction to a stimulus over a period of time and are similar to the ERPs. However, the ERSPs split the EEG signals into the diverse frequency bands to test whether a variation exists in the power of a specified frequency band over time [3].

Table 2. Common ERP components used in neuromarketing studies.

ERP	Description
P300	P300 is a positive potential that arises approximately 300 ms after the onset of a stimulus as a result of internal decision-making [16]. It shows the activity of attention in working memory as well as the adjustment to responses [17].
N200	N200 is a negative potential related to unfamiliarity, and reaches its peak between 200 and 350 ms after the onset of a stimulus [16].
LPC	The late positive component is a positive potential that arises approximately 400–800 ms after the onset of a stimulus associated with explicit recognition memory [18].
LPP	The late positive potential indicates enabled attention to emotional stimuli of either a positive or negative valence. This can be translated as neutral, pleasant, or unpleasant contextual stimuli.
N400	N400 is a negative potential related to oddness experiments such as brand names with products [10]
FN400	FN400 is located in the front-center of the brain related to familiarity and arises approximately 300–500 ms after the onset of a stimulus. FN400 has a more negative potential for new words than for similar and familiar words [18]
FRN	Feedback related negativity is a front-central negative potential related to a subject's choices. It arises in response to passively observed products 200–300 ms after the presentation of unfavorable versus favorable products [3].
PSW	Positive slow waves are correlated with sustained attention to visual emotional stimuli and can be detected long after the appearance of an emotional stimulus [10].

3. Literature Review

This section details EEG-based preference recognition, specifically, the neural correlation of the preference, predictive features of the preference, and preference classification algorithms.

Preference can be defined as a human attitude toward a collection of entities that can be mirrored in an explicit decision-making procedure. This aspect can also be an evaluative judgment in the sense of liking or disliking an object [19]. The possibility of measuring the conscious and unconscious brain activity in assessing advertisements, through processing the consumer's processing of the advertisement message, cognitive workload, and emotional state, cannot be disregarded. The idea of a 'buy button' in the brain may be overexaggerated; however, the research efforts to utilize the neural measures in monitoring the consumer thought processes are not trivial [20]. Understanding the neural process behind the preference, feelings, and decision-making can enhance the prediction of the user preferences and choices, and neuromarketing provides a precise objective determination of the implicit preferences of the consumers [21].

Several studies [10,22–24] have shown that the EEG can be used to determine the consumer preferences. To better utilize the EEG in consumer neuroscience research, the psychological processes underlying the consumer preferences must be understood.

In the following subsection, we describe the neural correlations of the EEG-based preference. Next, we classify the relevant studies into: (1) predictive features; and (2) classification algorithms of the preference recognition. Finally, we explain how the preferences can be detected using BCI.

3.1. Neural Correlations of the Preference

This subsection explains the neural elements correlated with the preferences. Certain areas of the brain are responsible for various cognitive and mental functions. To determine the positions of EEG electrodes, the underlying brain regions that are responsible for preference processing must be understood. Studies have demonstrated that the preference is linked to the frontal brain regions, specifically, the medial prefrontal cortex, nucleus accumbens [19,25], and medial orbitofrontal cortex [19,26].

Knutson et al. [25] linked the choice prediction of the products to the nucleus accumbens. When a consumer views the product, a higher activation of this region indicates a higher probability of the consumer purchasing that product. Furthermore, Kirk et al. [26] proved the relationship between the contextual preference and the medial orbitofrontal cortex; a higher activation in this region is related to higher level of preferences.

Recording the neural activity correlated with a certain function requires placing the electrodes directly above the corresponding brain area. Figure 2 shows the main electrode positions and the associated neural activity according to the 10-20 [27].

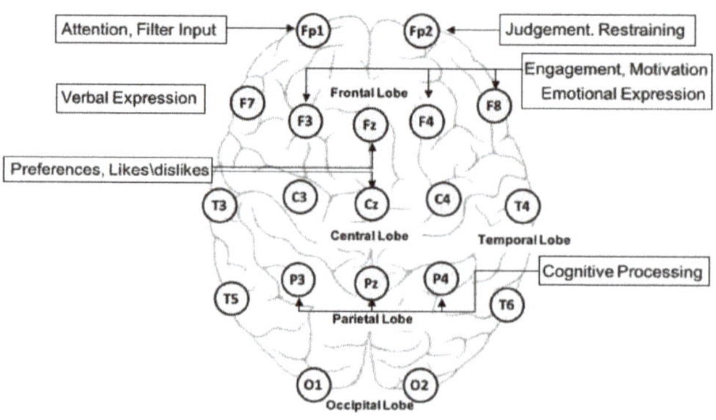

Figure 2. EEG electrodes and related neural functions.

Although many researchers [24–26] have proved that the medial-frontal cortex is responsible for the preference function, no consensus exists on which electrodes should be used within the same brain area. Table 3 summarizes the electrode positions used in the preference recognition research.

The authors of [24] proved that the medial-frontal cortex is linked to the individual preference in the beta range (16–18 Hz) on the mid-frontal areas on electrodes AFz, F2, FC1, and FCz. Moreover, the authors proved that the population preference is linked to the frontocentral areas in the theta range (60–100 Hz) on electrodes F1, F2, F4, FC3, FC1, FCz, FC2, FC4, C5, C3, C1, C4, and CP5.

Agarwal et al. [27] linked preference attributes such as attention, emotions, and liking to electrodes F3, C3, P3, Pz, Fz, Cz, and C4.

Vecchiato et al. [28] found that asymmetrical increments of the theta and alpha bands are linked to watching pleasant (unpleasant) advertisements, as noted in the left (right) brain areas at electrodes Fp1 (Fp2), AF7 (AF8), F7 (F8), and F1 (F2). The spectral power for alpha bands increases noticeably for liked advertisements at electrode F1 and for the disliked advertisements at electrodes AF8 and AF4. In the theta band, increased activity occurs at electrodes F2, AF8, and F3 for the disliked advertisements and at Fp1 for the liked advertisements.

Touchette et al. [29] found that the frontal asymmetry in the alpha band is linked to the consumers' unconscious reactions to the product attractiveness at electrodes F3 and F4. Vecchiato et al. [28] found that the asymmetrical frontal activity is statistically significantly positive in the alpha and theta bands between F1 and F2. In addition, this activity is significantly negative in the theta band between Fp2 and Fp1, AF8 and AF7, and F8 and F7.

Table 3. Common rhythms/ERP and electrode positions used for the preference detection in neuromarketing.

Reference	Rhythms	Electrode Channel
[24]	Beta range (16 – 18 Hz)	AFz, F2, FC1, and FCz.
[24]	Theta range (60 – 100 Hz)	F1, F2, F4, FC3, FC1, FCz, FC2, FC4, C5, C3, C1, C4, and CP5.
[28]	Theta	Fp1/Fp2, AF7/AF8, F7/F8, and F1/F2.
[28]	Alpha	Fp1/Fp2, AF7/AF8, F7/F8, and F1/F2.
[29]	Alpha	Fp3, and F4.
[27]	N200	F3, C3, P3, Pz, Fz, Cz, and C4.

EEG Indices

Based on our literature review, we identified four autonomic EEG indices that have been utilized for evaluating the reactions of the people in marketing stimuli: (1) the approach–withdrawal (AW) motivation index; (2) effort index; (3) choice index; and (4) valence. Such indices assist marketers in understanding customer responses to products [30,31].

1. **AW Index**

The AW index is also known as the frontal alpha asymmetry, which indicates motivation, desire, or approach avoidance. The frontal asymmetry theory, which was initiated in 1985, states that the frontal regions of the left and right hemispheres are responsible for positive feelings (approach motivation) and negative feelings (withdrawal motivation) [29], respectively. This index can be defined as the difference between the two hemispheres in the prefrontal alpha band, that is, the relative engagement of the frontal left hemisphere compared with the right one. Positive AW values correspond to positive motivation (approach behaviors), expressed in terms of the higher activation of the left frontal cortex. In contrast, negative AW values correspond to negative motivation (avoidance behaviors), expressed in terms of the higher activation of the left frontal cortex [29–32].

Numerous researchers have demonstrated the reliability and dependability of the frontal alpha asymmetry as an effective marker in the emotion and neuromarketing research [29–34]. Touchette [29] calculated the frontal alpha asymmetry scores by considering the difference between the right and left power spectral densities divided by their sum, as obtained using electrodes F4 and F3.

$$AW = \frac{alpha(F4) - alpha(F3)}{alpha(F4) + alpha(F3)} \tag{1}$$

2. **Effort Index**

   The effort index is defined as the frontal theta activity in the prefrontal cortex. A higher theta power in the frontal region has been linked to higher levels of task difficulty and complexity. This index acts as a sign of cognitive processing that results from mental fatigue [33], and it has been investigated extensively in neuromarketing research [3,24,28,33,35]. This factor demonstrates the importance of positive and negative emotional processing for the creation of the steady memory traces during advertising [30].

3. **Choice Index**

   The choice index is defined in terms of the frontal asymmetric gamma and beta oscillations, which are mostly linked to the real decision-making stage. It is also the most related element to willingness-to-pay responses, especially in the gamma band, for evaluating consumer preference and choice. Higher values of gamma and beta bands indicate a stronger activation of the left prefrontal region, and lower values are linked to relatively stronger activation of the right region [32]. Ramsoy et al. calculated the choice index for each band individually (gamma and beta) using electrodes AF3 and AF4 according to Equation (2):

$$\text{Choice index} = \frac{log(AF3) - log(AF4)}{log(AF3) + log(AF4)} \tag{2}$$

4. **Valence**

   Frontal asymmetry has been linked to preferences expressed as valence (i.e., the direction of a customer's emotional state). Left and right frontal activation is related to positive and negative valence, respectively. Numerous studies have supported the hypothesis that the frontal EEG asymmetry is an indicator of valence [34].

3.2. Predictive Features for the Preferences

This section reports on the studies that focused on the predictive features of neuroscience methods that can aid marketers in forecasting consumer preferences, as described in Table 4. Most of these studies employed distinctions of the standard regression analyses toward their prediction models. We classified these predictive features based on the EEG signal types: (1) rhythms; and (2) transient activities.

Table 4. EEG-based neuromarketing studies.

Year	Reference	Stimuli and Marketing Element	Signal Type	EEG Waves or ERP Component	Recording Modality
2008	[17]	Viewing brands and products	ERP	P300	BCI
2009	[11]	Watching TV ads	EEG	Alpha (for emotional state)	EMG, BCI and GSR
2010	[36]	Viewing brands and images	ERP	N200 and P300	BCI
2010	[37]	Watching TV ads	EEG	Alpha band frontal asymmetry	BCI
2010	[35]	Watching TV ads	EEG	Theta and gamma	Heart rate, BCI and GSR
2011	[28]	Watching TV ads	EEG	Asymmetrical increase in theta and alpha in PSD	BCI
2012	[38]	Viewing brand names	ERP	N400	BCI and EOG for eye movement
2012	[18]	Viewing products and prices	ERP	FN400, LPC, and P200	BCI
2013	[39]	Viewing products	EEG	Alpha, beta, theta, gamma, and delta	BCI, Eye tracking
2014	[10]	Viewing products	ERP	P300	BCI
2014	[41]	Watching TV ads	EEG	Theta and alpha	Heart rate, BCI and GSR
2015	[3]	Viewing products	ERSP and ERP	Theta, N200, and FRN	BCI
2015	[24]	Watching ads (movie trailers)	EEG (64 electrodes)	Beta and gamma oscillations	BCI and EOG for eye movement (2 electrode)
2016	[20]	Watching TV ads	dense-array EEG	Three epochs: 200-350, 350-500, and 500-800	BCI
2016	[42]	Viewing brand names	ERP	LPP	BCI
2017	[43]	Viewing product images	ERP	N200, LPP, and PSW	BCI
2017	[30]	Watching ad videos	EEG	Theta and alpha	Heart rate, BCI and GSR
2017	[22]	Viewing product images	EEG	Delta, theta, alpha, beta, and gamma	BCI
2017	[29]	Viewing product images	EEG	Alpha	BCI
2018	[44]	Viewing ads of food products	EEG	Delta, theta, alpha, beta, and gamma	BCI
2018	[32]	Viewing products and prices	EEG	Theta	BCI
2018	[31]	Tasting drinks	EEG	Alpha	BCI
2018	[33]	Viewing and touching products	EEG	Alpha and theta	BCI
2019	[27]	Viewing product images	ERSP, ERP	Theta, beta and N200	BCI
2019	[45]	Viewing tourism images, videos and words	EEG	Delta, theta, alpha, beta and gamma	BCI and GSR

3.2.1. Rhythms as Features

The beta and gamma oscillations from consumers who watched movie trailers were utilized to predict the box office sales and recall [24]. These factors were also used as an indicator of the willingness-to-pay to evaluate consumer preference and choice [32].

The alpha oscillations were used to compute the neural likeness and forecast recall and ticket sales. High-frequency EEG components were connected to both the individual preference (beta wave) and population preference (gamma wave) [24].

In addition, alpha frontal asymmetry was linked to the consumers' unconscious reactions to the product attractiveness [29]. Similarly, Modica et al. [33] linked the higher alpha frequencies to comfort food as well as foreign food products. Moreover, awarded campaigns (i.e., the campaigns that received prizes) in anti-smoking public service announcements were linked to higher alpha frequencies [30].

Lower theta frequencies were associated with the negative results toward choosing products [3]. Moreover, these frequencies have been linked to effective anti-smoking public service announcements [30] and foreign products compared with local products [33].

3.2.2. Transient Activities as Features

In the cognitive processes associated with preferences, several research studies considered the ERP components N400, N200, and P300, each of which can be described as follows.

1. **N400**

   Some researchers [10] found that the N400 component can reflect familiarity in forecasting hits in brand extension. A powerful association with well-known brand names was replicated in the case of larger N400 amplitudes, foreseeing greater consumer preference. Another brand extension study [38] reported that the N400 component is associated with the unconscious conceptual categorization of products and brands, albeit not with conscious assessments.

2. **N200**

   Some other researchers [36] observed that the N200 amplitude exposed a relationship between the emotional state and brand extension categories. This relationship appeared only with negative emotions and moderate brand extensions. A second study [43] suggested that N200 could indicate the product preferences, as determined by spontaneous procedures, whereas the LPP and PSW could indicate the product preferences, as determined by the conscious cognitive procedures. In a third study, the Cerebro system [27] combined the N200 mean, N200 minima, and ERSP to rank products according to customer preferences. Similarly, in a fourth study [3], the researchers used two methodologies, ERSP and ERP, to forecast the product preferences by examining theta brainwaves, N200, and FNR.

3. **P300**

   The consumer preferences for the expanded brand labels were clarified using greater P300 amplitudes [17]. The authors of [40] used P300 as a measure of the consumer preferences for certain product features.

   Other researchers [18,39] have considered factors that influence the purchasing decisions. The authors of [18] investigated the roles of mathematical ability, gender, pricing, and discount promotions in the process of consumer purchasing using the active BCI. The authors correlated the 'buy' decisions with ERP components, such as P200 and P300. To understand the product preferences, the authors also evaluated the relative importance (mutual information) of the diverse product (i.e., cracker) characteristics involved in the decision-making process by evaluating the cognitive processing by using the EEG alpha, beta, and theta brainwaves [39]. The researchers used eye tracking for choosing the preferred product. Michael et al. [45] used the same approach

(EEG with eye tracking) to investigate the emotional reactions of tourism preferences by using different stimuli (words, images, and video). The authors observed that the images had higher affective responses than those of words in travel decision-making driven by the unconscious preference.

The authors of [22] built a predictive model for consumer product choice from the EEG data. The researchers studied the roles of gender and age in the process of consumer preferences in terms of liking/disliking by using a passive BCI. Another research study [20] involved the use of an inductive research method to evaluate three successful and three unsuccessful advertisements by using a dense array EEG data. The results suggest that statistically significant ERP differences existed between the successful and unsuccessful advertisements.

3.3. Preference Classification Algorithms

Although considerable progress has been made in connecting the brain activities to the user choice, indications that neural assessment could genuinely be beneficial for forecasting the success of marketing activities remain limited [24]. The neural assessments can significantly increase the predictive power above and beyond that of the traditional assessments. Because the neural assessments are better predictors than self-reported assessments, the capability of neuroscience methods to forecast the preferences in real-world situations has incredible consequences for marketers. The first study to address this was published in 2007, and it was concluded that the pre-decisional activation in the related brain areas could be used to forecast the consequent choices [46]. Since then, many neuromarketing studies have published similar conclusions.

In recent years, it has become common practice to use multivariate methods, such as pattern classification, to predict choices. For example, a classification approach can be used to predict the out-of-sample choices from "non-choice' neural responses to different products. The resulting models, which were founded on basic neuroscience methods, are more reliable for predicting the new states and settings compared to traditional market methods, such as focus groups and questionnaires. Moreover, these methods are more likely to be scalable, providing marketers with a deeper understanding of consumers and crucial economic outcomes [46].

Preference modeling using data-mining approaches can be classified into three general signal fields: time, frequency, and a combination of time and frequency. Time-based preference modeling exploits the discovery of the ERPs, as discussed in Section 3.2.2. Frequency-based modeling is accomplished by understanding the features gained by performing power spectrum analyses by generating delta, theta, alpha, beta, and gamma frequency bands, as explained in Section 3.2.1. In addition, different frequency-based feature extraction methods can be used; for instance, common spatial patterns (CSPs) and spectral filters were used in the preference classification for music with an SVM, and an accuracy of 74.77% and 68.22%, respectively, was obtained [47]. Fast Fourier transform (FFT) as the feature extraction method was used, and the SVM obtained an accuracy of 82.14% [48]. In another study, the researchers used the FFT with the radial SVMs for the preference classification and obtained an accuracy of 75.44% [49]. The last preference model combines time and frequency by analyzing the power spectrum at the time intervals that cover the entire duration of the post stimuli interval to assess the brain signals. Several traditional data-mining algorithms have been applied to classify preferences, and the utilization of different time–frequency analysis (TF) approaches has been considered [15,23,50] to detect the user preferences for music. The use of KNN led to an accuracy of 86.5% and 83.34% with different TF approaches, namely the Hilbert–Huang Spectrum (HHS) and spectrogram, respectively [15]. In their extended study, Hadjidimitriou and Hadjileontiadis [51], using familiar music data, managed to obtain a considerably higher accuracy of 91.0%. Another work involved the performance of the music preference classification by using the TF approaches, namely, the discrete Fourier transform with a KNN, and an accuracy rate of 97.99% was achieved. The researchers could achieve a similar accuracy result when using the quadratic discriminant analysis (QDA) at 97.39% [52].

Most researchers applied variations of standard regression analysis to their prediction models. However, numerous techniques and methods have been developed to process EEG to determine the preference state of consumers by using classification algorithms. A review of some experimental neuromarketing articles and comparisons of computational approaches is presented in Table 5.

Table 5. Computational approaches for assessing the customer preferences.

	Method	Number of Studies	References
Preprocessing	Low-pass filter	3	[22,24,38]
	Band-pass filters	10	[3,11,15,20,28,39–41,44,53]
	Down sampling	3	[11,15,44]
	Average	7	[3,15,27,37,38,40,54]
Feature Extraction Selection	Statistical features: median, standard deviation, t-test, z-score transformation	9	[3,11,24,28,35,37,38,40,41]
	Time–frequency analysis	2	[15,54]
	Wavelet domain: entropy, energy » using wavelet transform	2	[22,39]
	Frequency domain: power spectral density, band power, spectrum power » using Fourier transform	8	[11,27,37,39,44,53–55]
	Correlation analysis: mutual information, correlation coefficient	3	[18,27,39]
Regression/Classification	Univariate linear regression analysis	2	[3,24]
	SVM	4	[15,22,54,55]
	KNN	3	[15,54,55]
	DNN	3	[22,23,55]
	RF	1	[22,55]
	HMM	1	[22]
	Linear regression (Lasso, Ridge)	1	[27]

Some preference studies involved the use of more than two classification algorithms to discover well-matched classifiers for a definite feature set [12]. Chew et al. [54] measured the user preferences for the aesthetics presented as virtual 3D shapes by using EEG. The researchers used the frequency bands as features to classify EEG into two classes—liked and disliked—by using the KNN and SVM and achieved high classification accuracies of 80% and 75%, respectively. However, these results cannot be considered reliable because the authors used an extremely small dataset of five subjects. In their extended study [55,56], the authors increased the number of subjects to 16 but better results were not obtained. Hakim et al. [44] achieved an accuracy of 68.5% by using the SVM to predict the most and least favored products by combining EEG measures with questionnaire measures.

Classifier combinations such as boosting, voting, or stacking can be used to join numerous classifiers, by merging their outputs and/or training them to complement each other and improve their performance [57]. The selection of the classification algorithms in a BCI system is mostly based on both the form of the acquired mental signals and the context in which the application is expected to be used. However, LDA and SVM are the most commonly applied classification algorithms and have been used in more than half of the EEG-based BCI articles.

Another categorization of the classifications was based on the survey research [57], which considered the BCI and machine learning literature from 2007 to 2017. The findings of the recently designed classification algorithms were divided into four main categories: adaptive classifiers, matrix and tensor classifiers, transfer learning, and deep learning. The adaptive classifiers are classifiers whose parameters, such as the feature weights, are gradually re-assessed and revised over time as new EEG data are presented. The matrix and tensor classifiers (multi-way array) avoid the use of the filters and feature selections and map the data directly onto a certain space with appropriate measures. The transfer learning approach aims to improve the performance of a learned classifier trained on a domain based on the information acquired, while learning another domain or task.

In recent times, deep learning has been employed in EEG-based preference recognition. DNNs are gatherings of the artificial neurons organized in layers to estimate the nonlinear resolution border. The most popular type of DNN used for BCIs is the multi-layer perceptron (MLP), which normally consists of only one or two hidden layers. Other DNN types have been explored less frequently, such as the Gaussian classifier neural networks or learning vector quantization neural networks [57]. Furthermore, Teo et al. [55,56] proposed deep learning approaches for preference recognition by using 3D rotating objects. The results prove that the use of the deep network could obtain a higher accuracy compared to that of the other machine learning classifiers, such as the SVM, RF, and KNN algorithms. In their extended research, Teo *et al.* [23] improved the result accuracies by using a deep network plus dropout architecture, with rectified linear units and tanh for activation at 79.76%.

Table 6 presents some neuromarketing studies that used different classification algorithms to obtain the most accurate results in predicting the consumer preferences. The review highlights the need to use more features and hybrid classifiers to improve the accuracy results of the predictions [22,44].

3.4. Preference Detection Using a BCI

This section explains the design process of the neuromarketing experiment to predict the consumer preferences and choices. First, a BCI device must be placed on the head of a consumer. Next, the consumer is asked to look at the products. During the recording phase, the EEG data are recorded concurrently while the consumer views a product. After viewing each product, the user is asked for his or her preference toward the product in terms of a five- or nine-point scale of subjective rank. When all products are displayed, the subjective ranks must be manually labeled as liked or disliked classes. Next, the EEG signals undergo a signal preprocessing and feature extraction. The classification module is developed based on the ground truth completed by the consumer's selection (subjective ranks).

Figure 3 presents a proposed BCI system for the preference detection composed of three main modules: signal preprocessing, feature extraction and selection, and classification.

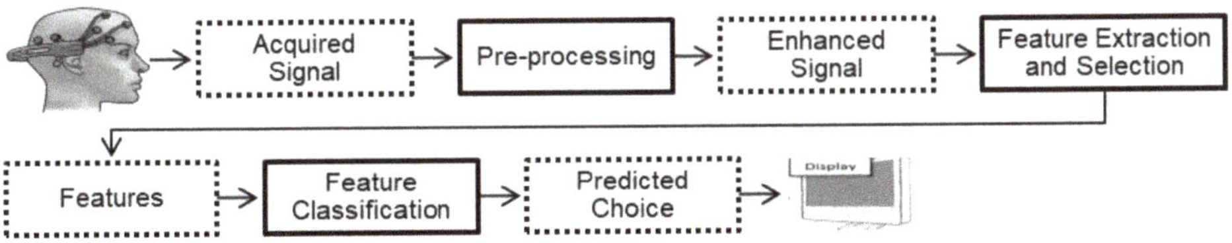

Figure 3. EEG-based consumer preference prediction system.

Table 6. Classification algorithms applied for recognizing the consumer preferences.

Reference	Classification Algorithm	Feature		Class	Best Accuracy
		Channel	Rhythms		
[54]	SVM	F3 and F4	(Alpha and theta), and (alpha, beta and theta)	1.Liked 2.Disliked	75%
	KNN	Fz, F3, F4	Alpha, theta and delta		80%
[49]	SVM	AF3, F7, F3, FC5, T7, P7, O1, O2, P8, T8, FC6, F4, F8, and AF4	Alpha, beta, theta, and gamma	1.Preferred image 2. Unnoticed image	75.44%
		F8 and T8	Alpha		83.64%
[15,51]	KNN + TF	AF3, F7, F3, FC5, T7, P7, O1, O2, P8, T8, FC6, F4, F8, and AF4	Delta, theta, alpha, beta and gamma	1.Liked 2.Disliked	83.34%
			Beta and gamma		86.5%
					91.02%
[47]	SVM + Spectral filter	FP1 and FP2	Delta, theta, alpha, beta and gamma	1.Liked 2.Disliked	68.22%
	SVM + CSP				74.77%
[52]	QDA + Fourier transform	AF3, AF4, F7, F8, F3, F4, FC5, FC6, T7, T8, P7, P8, O1, and O2	Theta, alpha, beta and gamma	1.Most preferred 2.Preferred 3.Less preferred 4.Least preferred	97.39%
	KNN +Fourier transform		Theta, alpha, beta, gamma and their corresponding symmetric differences		97.99%
[23,55,56]	DNN	POz, Fz, Cz, C3, C4, F3, F4, P3, and P4	Delta, theta, alpha, beta and gamma	1.Liked 2.Disliked	63.99%
	SVM				60.19%
	DNN with dropout and ReLu				79.76%
	DNN with dropout and tanh				74.38%
[44]	Logistic Regression	F7, Fp1, Fpz, Fp2, F8, Fz, Cz, and Pz	Delta, theta, alpha, beta and gamma	1.Most favored 2.Least favored	67.32%
	SVM				68.50%
	KNN				59.98%
	Decision trees with Adaboost M1				63.34%
[22]	DNN (2 layers + Sigmoid)	AF3, F7, F3, FC5, T7, P7, O1, O2, P8, T8, FC6, F4, F8, and AF4	Delta, theta, alpha, beta and gamma	1.Liked 2.Disliked	60.10%
	SVM				62.85%
	RF				68.41%
	HMM				70.33%

4. Proposed System for the EEG-based Preference Detection

The performance of EEG recognition systems is based on the selection of a feature extraction technique and a classification algorithm. In our study, we investigated the possibility of detecting two preference states, namely pleasant and unpleasant, by using EEG and classification algorithms. To this end, we performed rigorous offline analysis to investigate the computational intelligence for the preference detection and classification. We used deep learning classification from the DEAP dataset to explore how to employ intelligent computational methods in the form of classification algorithms. This could effectively mirror the preference states of the subjects. Furthermore, we compared our classification performance with those of the KNN and RF classifiers. We built our model in the open source programming language Python and used the Scikit-Learn toolbox for machine learning, along with SciPy for EEG filtering and preprocessing, MNE for EEG-specific signal processing, and the Keras library for deep learning.

In this section, we discuss our methodology along with some implementation details of the proposed system for EEG-based preference detection. We begin with describing the benchmark dataset and ground truth of the preference labeling. Next, we explain the feature extraction. Finally, we illustrate the DNN classification model.

4.1. Dataset Description

DEAP [58] is a benchmark EEG database developed for affective analysis. The DEAP database was built at the Queen Mary University in London, and it has been used in several research studies for preference detection [59,60]. Table 7 summarizes some characteristics of the DEAP dataset.

Table 7. DEAP dataset description.

Affective Model	Dimensional emotion model (Valence-Arousal-Dominance)
Stimuli	Visual- and audio-based stimuli (1-min music video)
Participants	32 participants aged 19 to 37 years
Trials	1280 trials (40 trials for each subject)
EEG device	32 EEG channels of the Biosemi Active Two system. The EEG data stream was collected using 32 Ag/AgCl electrodes, which were arranged in accordance with the 10-20 international system.
Experimental Protocol	Each participant watched and rated his or her emotional responses to 40 music videos on scales of arousal, valence, and dominance using self-assessment manikins (SAM). Participants also reported their liking of and familiarity with the videos.
DEAP Database	Different datasets that are publicly available in the DEAP database include recorded signal data, frontal face videos for a subset of participants, stimuli-volunteers' self-reported data, and subjects' self-assessments.

4.2. Preference Modeling and Ground Truth

To set the true preferences (ground truth table), we used the DEAP self-assessment reports to identify the preference states using a nine-point Likert scale for valence dimension. In this study, we considered the valence dimension as a preference indicator to align with the target preference state: pleasant and unpleasant. Moreover, we considered EEG trials that had at least two different valence levels—low and high. The valence levels are classified as follows: (1) low valence if the valence rating is between 1 and 5; and (2) high valence if the valence rating is between 6 and 9. The presence of a low or high valence is an indicator of an unpleasant or pleasant preference state, respectively.

4.3. Data Pre-Processing

We used the preprocessed EEG dataset from the DEAP database, where the sampling rate of the original recorded data of 512 Hz was down-sampled to a sampling rate of 128 Hz, with a bandpass frequency filter that ranged from 4.0 to 45.0 Hz, and the EOG artifacts were eliminated from the signals using a blind source separation method, namely independent component analysis ICA. The data were averaged and segmented to 60-s trials. Then, we applied a channel selection step with a dimensionality reduction technique. The aim of this step was to reduce the number of features and/or channels used by selecting a subset that excludes very high-dimensional and noisy data. Ideally, the features that are meaningful or useful in the classification stage are identified and selected, while others, including outliers and artifacts, are omitted. Moreover, it reduces the computational cost of the subsequent steps. Therefore, we keep only the channels in which we are interested (Fz, AF4, AF3, F4, and F3).

4.4. Feature Extraction

Feature extraction plays a crucial role in building EEG-based BCI applications. Thus, we extracted the EEG frequency bands by using a power spectral density (PSD) method called the Welch method. Subsequently, we used the resulting frequency bands to calculate the valence as a preference indicator. Figure 4 presents the block diagram of feature extraction.

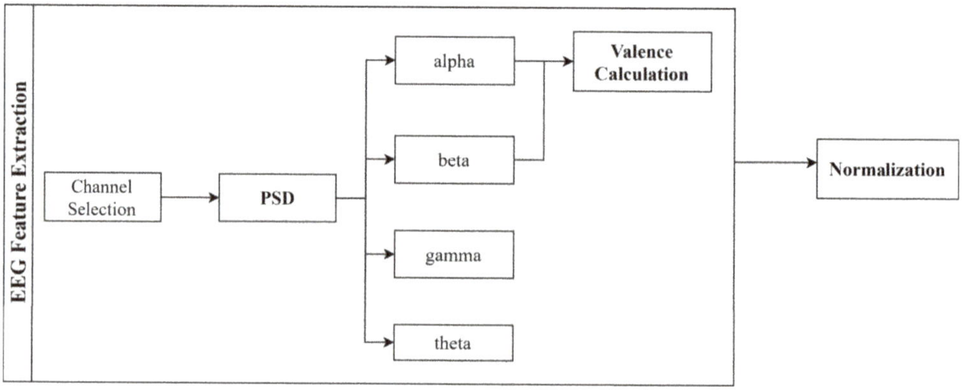

Figure 4. Feature extraction block diagram.

4.4.1. PSD

The PSD is one of the most popular feature extraction methods based on the frequency domain analysis in the neuromarketing research. Research studies [11,37,39] have demonstrated that the PSD obtained from the EEG signals works well for determining consumer preferences. The PSD method converts the data in the time domain to the frequency domain and vice versa. This conversation is based on the FFT, which calculates the discrete Fourier transform and its inverse.

We used the PSD technique in this study to divide each EEG signal into four different frequency bands: theta (4–8 Hz), alpha (8–13 Hz), beta (13–30 Hz), and gamma (30–40 Hz). The Python signal processing toolbox (MNE) was used for PSD calculation, and the average power over the frequency bands was computed to build a feature using the avgpower function in the MNE toolbox.

4.4.2. Valence

The valence was selected as the measure of preference in this study. Strong valence is reflected in the activation of frontal EEG asymmetry [34]. In DEAP dataset [58], there was high correlation between valence and EEG frequency bands, as shown in Figure 5. The increment in valence led to power increment in alpha, which is consistent with the results in a similar study [34]. We did not use liking rating in the DEAP dataset because the data owners [58] reported conflicting findings between the activation in left alpha power and liking.

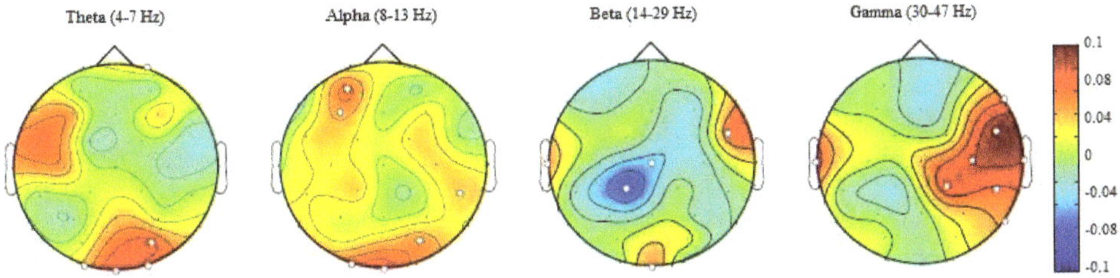

Figure 5. The average correlations for all subjects of the valence ratings with the power of different frequency bands. The highlighted electrodes correlate significantly ($p < 0.05$) with the valence ratings. © [2011] IEEE. Reprinted, with permission, from:*IEEE Trans. Affect. Comput., DEAP: A Database for Emotion Analysis Using Physiological Signals, Mühl, C.; Lee, J.s. [58].*

We applied the different valence equations and investigated the relationship with the DEAP self-assessment valence measurement. For the valence calculation, we used the extracted alpha and beta band powers from the DEAP data and considered only the following electrodes: Fz, AF3, F3, AF4, and F4. Finally, we computed the values of the valence by using four different equation (Equations (3)–(6)), which have been well-explained in a previous paper [34] authored by an author of this paper.

$$\text{Valence} = \frac{beta(AF3, F3)}{alpha(AF3, F3)} - \frac{beta(AF4, F4)}{alpha(AF4, F4)} \tag{3}$$

$$\text{Valence} = ln[alpha(Fz, AF3, F3)] - ln[alpha(Fz, AF4, F4)] \tag{4}$$

$$\text{Valence} = alpha(F4) - beta(F3) \tag{5}$$

$$\text{Valence} = \frac{alpha(F4)}{beta(F4)} - \frac{alpha(F3)}{\text{beta}(F3)} \tag{6}$$

4.5. DNN Classification

Deep learning has been proved as an effective tool to help make the EEG signals meaningful because of its ability to learn the feature representations from the raw data. DNNs are models consisting of the combined layers of "neurons" in which each layer applies a linear transformation to the input data. Then, the transformation result of each layer undergoes processing on the basis of a nonlinear cost function. The parameters of such transformations are deduced by minimizing a cost function [61]. The DNN operates in one forward direction, from the input neurons through the hidden ones (if available) to the output neurons in the forward directions. Assuming that the length window of the samples is s, the input of the DNN for the EEG signals consists of a multidimensional array $X_i \in \mathbb{R}^{e \times s}$ that contains s samples associated with a window for all e electrodes. The fully connected layer, which is the most common type of layer used in building a DNN, consists of fully connected neurons. The input of every neuron is the activation of each neuron from the previous layer [61].

Our study aimed to detect two preference states in the EEG data. Therefore, we employed intelligent classification algorithms that could effectively mirror the preferences of the subjects. We proposed a DNN classifier and compared its performance with those of KNN and RF classifiers.

The block diagram of the proposed DNN classifier is shown in Figure 6. First, the extracted features are normalized using minimum–maximum normalization (Equation (7)) and then fed into the DNN classifier.

$$x\_\text{scaled} = (x - \text{min}) / (\text{max} - \text{min}) \tag{7}$$

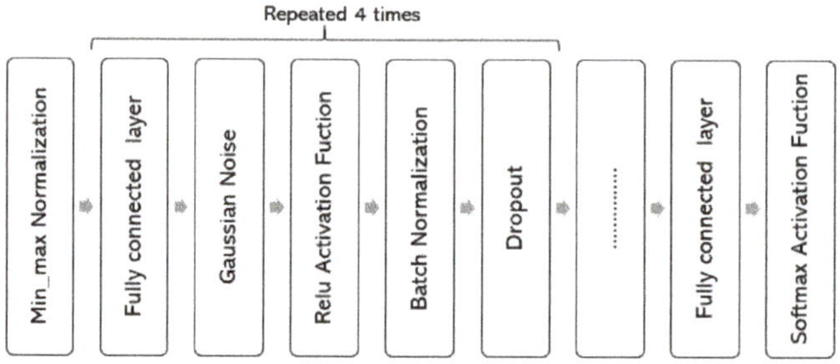

Figure 6. Block diagram of the DNN classifier.

In this study, the considered DNN architecture is a fully connected feed-forward neural network with three hidden layers, which contain units involving rectified linear activation functions (ReLu). The output is obtained as a soft-max layer with a binary cross-entropy cost function. The input layer consists of 2367 units, and each hidden layer consists of 75% units from its predecessor (previous) layer. In particular, the first, second, and third hidden layers involve 1800, 1300, and 800 units, respectively. The output layer dimensions pertain to the number of target preferences state (2) units. To train the DNN classifier, we used Adam gradient descent with three objective loss functions: binary cross-entropy, categorical cross-entropy, and hinge cross function. For transfer learning, we considered the reasonable defaults and followed the established best practices: the start learning rate was 0.001. Then, we linearly reduced the rate with each epoch such that the learning rate for the last epoch was 0.0001. We set the dropout for the input and hidden layers as 0.1 and 0.05, respectively. The stopping criterion of the network training was determined according to the model performance on a testing set. If the network started to over-fit, the network training was stopped. This stopping criterion is helpful for reducing the possibility of over-fitting of the validation data. The network was tested on a test set, which contained approximately 20% of the data samples in the dataset.

5. Results and Discussion

We predicted the preference states (pleasant or unpleasant) using different classification algorithms: DNN, RF, KNN, and SVM. We used different evaluation measurements: accuracy, recall, and precision. The accuracy was calculated as the average of the binary measurements in which the score of every class was weighted by its availability in the real data. Precision is the proportion of pleasant preference predictions that were actually correct. Recall is the proportion of actual pleasant preferences that were successfully predicted. To evaluate the performance of classification algorithms, we used different cross validation methods: holdout (train/test splitting), k-folds cross validation, and leave-one-out cross validation (LOOCV). Table 8 presents the accuracy results of DNN, RF, KNN, and SVM for each cross validation method. In LOOCV, RF reached the best accuracy results at 90% while DNN reached similar results to RF at 93% in the holdout validation method. In k-fold validation method, KNN achieved the best accuracy results at 90% and 91% when k was set to 10 and 20, respectively. Because the best accuracy results were achieved using the holdout validation, this method was chosen as the base validation for comparison and the tuning the DNN hyper parameter (loss function).

The proposed DNN model was compared with three conventional classification algorithms for EEG signals: SVM, RF, and KNN. Table 9 presents the accuracy, recall, and precision results of RF, KNN, and DNN using three different loss functions in the DNN: the categorical cross-entropy function, binary cross-entropy function, and hinge function. The KNN classifier led to a better accuracy of 88% when K was set to 1. Although the RF achieved a high accuracy of 92%, the DNN reached the highest accuracy result of 94% with hinge cross-entropy function compared to the other conventional classification algorithms. To ensure that the DNN does not have over-fitting problem, we presented the loss per epoch for each cross-entropy function. The average loss per epoch DNN with the categorical, binary, and hinge function reached a value of 0.28, 0.24, and 0.23, respectively, as shown in Figure 7.

Table 8. Accuracy results of preference detection with DNN, RF, KNN, and SVM using different cross validation methods.

Cross Validation Method	LOOCV	Holdout	k-fold		
			(K = 5)	(K = 10)	(K = 20)
DNN (Binary Cross-entropy)	73%	93%	60%	68%	71%
SVM	47%	62%	49%	48%	47%
RF	90%	93%	86%	87%	89%
KNN	92%	88%	88%	90%	91%

Table 9. Results of preference recognition using holdout validation and different classifiers: DNN, SVM, RF, and KNN.

Classifier / Metrics	DNN			SVM	RF	KNN		
	Hinge Cross	Categorical Cross	Binary cross			K = 5	K = 3	K = 1
Accuracy	94%	91%	93%	62%	92%	73%	79%	88%
Recall	94%	91%	93%	62%	92%	73%	79%	88%
Precision	94%	92%	94%	64%	93%	75%	81%	90%

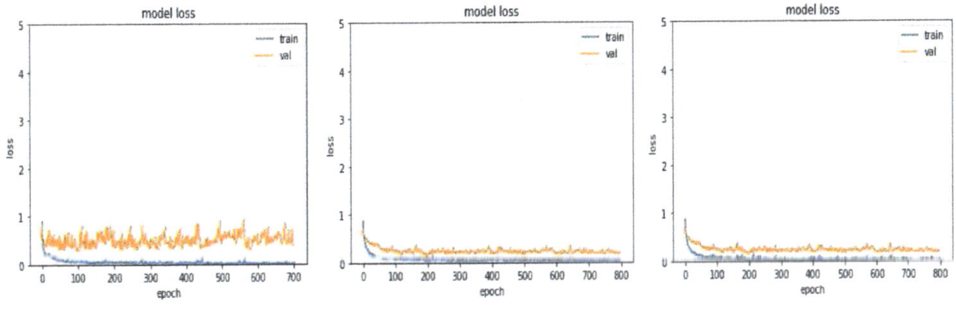

(a) (b) (c)

Figure 7. Loss per epoch on the training and validation sets in DNN using different cross-entropy functions: (**a**) categorical cross-entropy (average loss rate = 0.28); (**b**) Binary cross-entropy (average loss rate = 0.24); and (**c**) hinge cross-entropy (average loss rate = 0.23).

6. Conclusions

This study proposed a DNN model to detect the preferences from the EEG signals by using the pre-processed DEAP dataset. Two types of features were extracted from the EEG: the PSD and valence. This aspect resulted in a group of 2367 unique features illustrating the EEG activity in each trial. We used different evaluations measures (accuracy, recall, and precision) and various validation methods (holdout, LOOCV, and k-fold cross validation) to test classifiers' performance. We built four different classifiers, namely the DNN, RF, SVM, and KNN classifiers, which achieved an accuracy of 94%, 92%, 62%, and 88%, respectively. The results demonstrate that, although the proposed DNN exhibits a higher accuracy, recall, and precision compared with the KNN and SVM, RF reaches similar results to DNN on the same dataset. Future research directions will involve exploring the DNNs in the context of transfer learning for preference detection.

Author Contributions: M.A. conceived, designed, and performed the experiment; analyzed and interpreted the data; and drafted the manuscript. A.A.-N. co-supervised the analysis, reviewed the manuscript, and contributed to the discussion. M.Y. supervised this study. All authors have read and approved the submitted version of the manuscript.

Acknowledgments: The authors would like to thank the deanship of scientific research for funding and supporting this research through the initiative of DSR Graduate Students Research Support (GSR) at King Saud University.

Conflicts of Interest: The authors declare no conflict of interest.

Abbreviations

The following abbreviations are used in this manuscript:

BCI	Brain Computer Interface
EOG	Electrooculography
EEG	Electroencephalography
ERSP	Event Related Spectral Perturbations
PSD	Power Spectral Density
ICA	Independent Component Analysis
ERP	Event Related Potential
DNN	Deep Neural Network
SVM	Support Vector Machine
AW	Approach-Withdrawal
KNN	k-Nearest Neighbor
RF	Random Forest
LOOCV	Leave-One-Out Cross Validation
PCA	Principal Component Analysis
ReLu	Rectified Linear Unit
LPP	Late Positive Component
FRN	Feedback Related Negativity
MLP	Multi-Layer Perceptron
GSR	Galvanic Skin Response
HMM	Hidden Markov Model
LDA	Linear Discriminant Analysis
EMG	Facial electromyography
TF	time–frequency analysis
CSP	Common Spatial Pattern
FFT	Fast Fourier Transform
HHS	Hilbert–Huang Spectrum

References

1. Morin, C. Neuromarketing: The New Science of Consumer Behavior. *Society* **2011**, *48*, 131–135. [CrossRef]
2. Ait Hammou, K.; Galib, M.H.; Melloul, J. The Contributions of Neuromarketing in Marketing Research. *J. Manag. Res.* **2013**, *5*, 20. [CrossRef]
3. Telpaz, A.; Webb, R.; Levy, D.J. Using EEG to Predict Consumers' Future Choices. *J. Mark. Res.* **2015**, *52*, 511–529. [CrossRef]
4. Barros, R.Q.; Tavares, A.S.; Albuquerque, W.; da Silva, J.C.; de Lemos, I.A.; de Albuquerque Cardoso, R.L.S.; Soares, M.M.; Cairrao, M.R. *Analysis of Product Use by Means of Eye Tracking and EEG: A Study of Neuroergonomics*; Lecture Notes in Computer Science; Springer International Publishing: Cham, Switzerland, 2016; Volume 9747, pp. 539–548. [CrossRef]
5. Abdulkader, S.N.; Atia, A.; Mostafa, M.S.M. Brain computer interfacing: Applications and challenges. *Egypt. Inform. J.* **2015**, *16*, 213–230. [CrossRef]
6. Ramadan, R.A.; Refat, S.; Elshahed, M.A.; Ali, R.A. Brain-Computer Interfaces. In *Intelligent Systems Reference Library*; Springer International Publishing: Cham, Switzerland, 2015; Volume 74, pp. 31–50. [CrossRef]

7. Ramadan, R.A.; Vasilakos, A.V. Brain Computer Interface: Control Signals Review. *Neurocomputing* **2016**, *223*, 1–19. [CrossRef]
8. Agarwal, S.; Dutta, T. Neuromarketing and consumer neuroscience: current understanding and the way forward. *Decision* **2015**, *42*, 457–462. [CrossRef]
9. Murugappan, M.; Murugappan, S.; Balaganapathy; Gerard, C. Wireless EEG signals based Neuromarketing system using Fast Fourier Transform (FFT). In Proceedings of the 2014 IEEE 10th International Colloquium on Signal Processing and its Applications, Kuala Lumpur, Malaysia, 7–9 March 2014, pp. 25–30. [CrossRef]
10. Lin, M.H.J.; Cross, S.N.N.; Jones, W.J.; Childers, T.L. Applying EEG in consumer neuroscience. *Eur. J. Mark.* **2018**, *52*, 66–91. [CrossRef]
11. Ohme, R.; Reykowska, D.; Wiener, D.; Choromanska, A. Analysis of Neurophysiological Reactions to Advertising Stimuli by Means of EEG and Galvanic Skin Response Measures. *J. Neurosci. Psychol. Econ.* **2009**, *2*, 21–31. [CrossRef]
12. Hwang, H.J.; Kim, S.; Choi, S.; Im, C.H. EEG-Based Brain-Computer Interfaces: A Thorough Literature Survey. *Int. J. Hum.-Comput. Interact.* **2013**, *29*, 814–826. [CrossRef]
13. Krampe, C.; Gier, N.R.; Kenning, P. The Application of Mobile fNIRS in Marketing Research—Detecting the "First-Choice-Brand" Effect. *Front. Hum. Neurosci.* **2018**, *12*. [CrossRef]
14. Meyerding, S.G.; Mehlhose, C.M. Can neuromarketing add value to the traditional marketing research? An exemplary experiment with functional near-infrared spectroscopy (fNIRS). *J. Bus. Res.* **2020**, *107*, 172–185. [CrossRef]
15. Hadjidimitriou, S.K.; Hadjileontiadis, L.J. Toward an EEG-based recognition of music liking using time-frequency analysis. *IEEE Trans. Biomed. Eng.* **2012**, *59*, 3498–3510. [CrossRef] [PubMed]
16. Nezamfar, H. FlashLife™, a Context-Aware Code-VEP Based Brain Computer Interface for Daily Life Using EEG Signals. Ph.D. Thesis, Northeastern University, Boston, MA, USA, 2016.
17. Ma, Q.; Wang, X.; Shu, L.; Dai, S. P300 and categorization in brand extension. *Neurosci. Lett.* **2008**, *431*, 57–61. [CrossRef] [PubMed]
18. Jones, W.J.; Childers, T.L.; Jiang, Y. The shopping brain: Math anxiety modulates brain responses to buying decisions. *Biol. Psychol.* **2012**, *89*, 201–213. [CrossRef] [PubMed]
19. Ramsøy, T.Z.; Friis-Olivarius, M.; Jacobsen, C.; Jensen, S.B.; Skov, M. Effects of perceptual uncertainty on arousal and preference across different visual domains. *J. Neurosci. Psychol. Econ.* **2012**, *5*, 212–226. [CrossRef]
20. Daugherty, T.; Hoffman, E.; Kennedy, K. Research in reverse: Ad testing using an inductive consumer neuroscience approach. *J. Bus. Res.* **2016**, *69*, 3168–3176. [CrossRef]
21. Alvino, L.; Constantinides, E.; Franco, M. Towards a Better Understanding of Consumer Behavior: Marginal Utility as a Parameter in Neuromarketing Research. *Int. J. Mark. Stud.* **2018**, *10*, 90. [CrossRef]
22. Yadava, M.; Kumar, P.; Saini, R.; Roy, P.P.; Dogra, D.P. Analysis of EEG signals and its application to neuromarketing. *Multimed. Tools Appl.* **2017**, *76*, 19087–19111. [CrossRef]
23. Teo, J.; Chew, L.H.; Chia, J.T.; Mountstephens, J. Classification of Affective States via EEG and Deep Learning. *Int. J. Adv.. Comput. Sci. Appl.* **2018**, *9*, 132–142. [CrossRef]
24. Boksem, M.A.S.; Smidts, A. Brain Responses to Movie Trailers Predict Individual Preferences for Movies and Their Population-Wide Commercial Success. *J. Mark. Res.* **2015**, *52*, 482–492. [CrossRef]
25. Knutson, B.; Rick, S.; Wimmer, G.E.; Prelec, D.; Loewenstein, G. Neural Predictors of Purchases. *Neuron* **2007**, *53*, 147–156. [CrossRef] [PubMed]
26. Kirk, U.; Skov, M.; Hulme, O.; Christensen, M.S.; Zeki, S. Modulation of aesthetic value by semantic context: An fMRI study. *NeuroImage* **2009**, *44*, 1125–1132. [CrossRef] [PubMed]
27. Agarwal, M.; Sivakumar, R. Cerebro: A Wearable Solution to Detect and Track User Preferences using Brainwaves. In Proceedings of the 5th ACM Workshop on Wearable Systems and Applications, Seoul, Korea, 12 June 2019; pp. 47–52. [CrossRef]
28. Vecchiato, G.; Toppi, J.; Astolfi, L.; Fallani, F.D.V.; Cincotti, F.; Mattia, D.; Bez, F.; Babiloni, F. Spectral EEG frontal asymmetries correlate with the experienced pleasantness of TV commercial advertisements. *Med. Biol. Eng. Comput.* **2011**, *49*, 579–583. [CrossRef] [PubMed]
29. Touchette, B.; Lee, S.E. Measuring Neural Responses to Apparel Product Attractiveness: An Application of Frontal Asymmetry Theory. *Cloth. Text. Res. J.* **2017**, *35*, 3–15. [CrossRef]

30. Cartocci, G.; Caratu, M.; Modica, E.; Maglione, A.G.; Rossi, D.; Cherubino, P.; Babiloni, F. Electroencephalographic, Heart Rate, and Galvanic Skin Response Assessment for an Advertising Perception Study: Application to Antismoking Public Service Announcements. *Jove-J. Vis. Exp.* **2017**. doi:10.3791/55872. [CrossRef]
31. Cherubino, P. Application of Neuro- Marketing techniques to the wine tasting experience. In Proceedings of the 11th Annual Conference of the EuroMed Academy of Business, Valletta, Malta, 12–14 September 2018; pp. 290–298.
32. Ramsøy, T.Z.; Skov, M.; Christensen, M.K.; Stahlhut, C. Frontal brain asymmetry and willingness to pay. *Front. Neurosci.* **2018**, *12*. [CrossRef]
33. Modica, E.; Cartocci, G.; Rossi, D.; Martinez Levy, A.C.; Cherubino, P.; Maglione, A.G.; Di Flumeri, G.; Mancini, M.; Montanari, M.; Perrotta, D.; et al. Neurophysiological responses to different product experiences. *Comput. Intell. Neurosci.* **2018**, *2018*. [CrossRef]
34. Al-Nafjan, A.; Hosny, M.; Al-Wabil, A.; Al-Ohali, Y. Classification of Human Emotions from Electroencephalogram (EEG) Signal using Deep Neural Network. *Int. J. Adv. Comput. Sci. Appl.* **2017**, *8*, 419–425. [CrossRef]
35. Vecchiato, G.; Astolfi, L.; Fallani, F.D.V.; Cincotti, F.; Mattia, D.; Salinari, S.; Soranzo, R.; Babiloni, F. Changes in brain activity during the observation of TV commercials by using EEG, GSR and HR measurements. *Brain Topogr.* **2010**, *23*, 165–179. [CrossRef]
36. Ma, Q.; Wang, K.; Wang, X.; Wang, C.; Wang, L. The influence of negative emotion on brand extension as reflected by the change of N2: A preliminary study. *Neurosci. Lett.* **2010**, *485*, 237–240. [CrossRef]
37. Ohme, R.; Reykowska, D.; Wiener, D.; Choromanska, A. Application of frontal EEG asymmetry to advertising research. *J. Econ. Psychol.* **2010**, *31*, 785–793. [CrossRef]
38. Wang, X.; Ma, Q.; Wang, C. N400 as an index of uncontrolled categorization processing in brand extension. *Neurosci. Lett.* **2012**, *525*, 76–81. [CrossRef] [PubMed]
39. Khushaba, R.N.; Wise, C.; Kodagoda, S.; Louviere, J.; Kahn, B.E.; Townsend, C. Consumer neuroscience: Assessing the brain response to marketing stimuli using electroencephalogram (EEG) and eye tracking. *Expert Syst. Appl.* **2013**, *40*, 3803–3812. [CrossRef]
40. Wang, J.; Han, W. The impact of perceived quality on online buying decisions: An event-related potentials perspective. *Neuroreport* **2014**, *25*, 1091–1098. [CrossRef]
41. Vecchiato, G.; Maglione, A.G.; Cherubino, P.; Wasikowska, B.; Wawrzyniak, A.; Latuszynska, A.; Latuszynska, M.; Nermend, K.; Graziani, I.; Leucci, M.R.; et al. Neurophysiological Tools to Investigate Consumer's Gender Differences during the Observation of TV Commercials. *Comput. Math. Methods Med.* **2014**. [CrossRef]
42. Bosshard, S.S.; Bourke, J.D.; Kunaharan, S.; Koller, M.; Walla, P. Established liked versus disliked brands: Brain activity, implicit associations and explicit responses. *Cogent Psychol.* **2016**, *3*. [CrossRef]
43. Goto, N.; Mushtaq, F.; Shee, D.; Lim, X.L.; Mortazavi, M.; Watabe, M.; Schaefer, A. Neural signals of selective attention are modulated by subjective preferences and buying decisions in a virtual shopping task. *Biol. Psychol.* **2017**, *128*, 11–20. [CrossRef]
44. Hakim, A.; Klorfeld, S.; Sela, T.; Friedman, D.; Shabat-Simon, M.; Levy, D.J. Pathways to Consumers Minds: Using Machine Learning and Multiple EEG Metrics to Increase Preference Prediction Above and Beyond Traditional Measurements. *bioRxiv* **2018**. [CrossRef]
45. Michael, I.; Ramsoy, T.; Stephens, M.; Kotsi, F. A study of unconscious emotional and cognitive responses to tourism images using a neuroscience method. *J. Islamic Mark.* **2019**, *10*, 543–564. [CrossRef]
46. Plassmann, H.; Venkatraman, V.; Huettel, S.; Yoon, C. Consumer Neuroscience: Applications, Challenges, and Possible Solutions. *J. Mark. Res.* **2015**, *52*, 427–435. [CrossRef]
47. Pan, Y.; Guan, C.; Yu, J.; Ang, K.K.; Chan, T.E. Common frequency pattern for music preference identification using frontal EEG. *Int. IEEE/EMBS Conf. Neural Eng. NER* **2013**, 505–508. [CrossRef]
48. Tseng, K.C.; Lin, B.S.; Han, C.M.; Wang, P.S. Emotion recognition of EEG underlying favourite music by support vector machine. In Proceedings of the ICOT 2013—1st International Conference on Orange Technologies, Tainan, Taiwan, 12–16 March 2013; pp. 155–158. [CrossRef]
49. Kim, Y.; Kang, K.; Lee, H.; Bae, C. *Preference Measurement Using User Response Electroencephalogram*; Springer: Berlin/Heidelberg, Germany, 2015; pp. 1315–1324. [CrossRef]

50. Cohrdes, C.; Wrzus, C.; Frisch, S.; Riediger, M. Tune yourself in: Valence and arousal preferences in music-listening choices from adolescence to old age. *Dev. Psychol.* **2017**, *53*, 1777–1794. [CrossRef] [PubMed]
51. Hadjidimitriou, S.K.; Hadjileontiadis, L.J. EEG-Based classification of music appraisal responses using time-frequency analysis and familiarity ratings. *IEEE Trans. Affect. Comput.* **2013**, *4*, 161–172. [CrossRef]
52. Moon, J.; Kim, Y.; Lee, H.; Bae, C.; Yoon, W.C. Extraction of user preference for video stimuli using eeg-based user responses. *ETRI J.* **2013**, *35*, 1105–1114. [CrossRef]
53. Bercik, J.; Horska, E.; Wang, R.W.Y.; Chen, Y.C. The impact of parameters of store illumination on food shopper response. *Appetite* **2016**, *106*, 101–109. [CrossRef] [PubMed]
54. Chew, L.H.; Teo, J.; Mountstephens, J. Aesthetic preference recognition of 3D shapes using EEG. *Cognit. Neurodyn.* **2016**, *10*, 165–173. [CrossRef]
55. Teo, J.; Hou, C.L.; Mountstephens, J. Deep learning for EEG-Based preference classification. *AIP Conf. Proc.* **2017**, *1891*. [CrossRef]
56. Teo, J.; Hou, C.L.; Mountstephens, J. Preference classification using Electroencephalography (EEG) and deep learning. *J. Telecommun. Electron. Comput. Eng.* **2018**, *10*, 87–91.
57. Lotte, F.; Bougrain, L.; Cichocki, A.; Clerc, M.; Congedo, M.; Rakotomamonjy, A.; Yger, F. A review of classification algorithms for EEG-based brain-computer interfaces: A 10 year update. *J. Neural Eng.* **2018**, *15*, aab2f2. [CrossRef]
58. Mühl, C.; Lee, J.s. DEAP: A Database for Emotion Analysis Using Physiological Signals. *IEEE Trans. Affect. Comput.* **2011**. *3*, 18–31. [CrossRef]
59. Shin, S.; Jang, S.J.; Lee, D.; Park, U.; Kim, J.H. Brainwave-based mood classification using regularized common spatial pattern filter. *KSII Trans. Internet Inf. Syst.* **2016**, *10*, 807–824. [CrossRef]
60. Lan, Z.; Liu, Y.; Sourina, O.; Wang, L. Real-time EEG-based user's valence monitoring. In Proceedings of the 2015 10th International Conference on Information, Communications and Signal Processing (ICICS), Singapore, 2–4 December 2015; pp. 1–5. [CrossRef]
61. Roy, Y.; Banville, H.; Albuquerque, I.; Gramfort, A.; Falk, T.H.; Faubert, J. Deep learning-based electroencephalography analysis: A systematic review. *J. Neural Eng.* **2019**, *16*, 051001. [CrossRef] [PubMed]

MDPI
St. Alban-Anlage 66
4052 Basel
Switzerland
Tel. +41 61 683 77 34
Fax +41 61 302 89 18
www.mdpi.com

Applied Sciences Editorial Office
E-mail: applsci@mdpi.com
www.mdpi.com/journal/applsci

www.ingramcontent.com/pod-product-compliance
Lightning Source LLC
Chambersburg PA
CBHW061549210926
51850CB00009B/2053
9783036518022